AIoT and Smart Sensing Technologies for Smart Devices

Fadi Al-Turjman
AI and Robotics Institute, Near East University, Nicosia, Turkey & Faculty of Engineering, University of Kyrenia, Kyrenia, Turkey

A volume in the Advances in Computational Intelligence and Robotics (ACIR) Book Series

IGI Global
PUBLISHER of TIMELY KNOWLEDGE

Published in the United States of America by
IGI Global
Engineering Science Reference (an imprint of IGI Global)
701 E. Chocolate Avenue
Hershey PA, USA 17033
Tel: 717-533-8845
Fax: 717-533-8661
E-mail: cust@igi-global.com
Web site: http://www.igi-global.com

Library of Congress Cataloging-in-Publication Data

Names: Alturjman, Fadi, 1982- editor.
Title: AIoT and smart sensing technologies for smart devices / edited by Fadi Alturjman.
Description: Hershey, PA : Engineering Science Reference, [2024] | Includes bibliographical references and index. | Summary: "This book exchanges ideas on the advancements in smart technology, application areas for advanced communication systems and development of new services, facilitate a tremendous growth of new devices and smart things that need to be connected to the Internet through a variety of wireless technologies"-- Provided by publisher.
Identifiers: LCCN 2023037185 (print) | LCCN 2023037186 (ebook) | ISBN 9798369307861 (hardcover) | ISBN 9798369307878 (ebook)
Subjects: LCSH: Internet of things--Industrial applications.
Classification: LCC TK5105.8857 .A46 2023 (print) | LCC TK5105.8857 (ebook) | DDC 004.67/8--dc23/eng/20240118
LC record available at https://lccn.loc.gov/2023037185
LC ebook record available at https://lccn.loc.gov/2023037186

This book is published in the IGI Global book series Advances in Computational Intelligence and Robotics (ACIR) (ISSN: 2327-0411; eISSN: 2327-042X)

British Cataloguing in Publication Data
A Cataloguing in Publication record for this book is available from the British Library.

For electronic access to this publication, please contact: eresources@igi-global.com.

Advances in Computational Intelligence and Robotics (ACIR) Book Series

ISSN:2327-0411
EISSN:2327-042X

MISSION

While intelligence is traditionally a term applied to humans and human cognition, technology has progressed in such a way to allow for the development of intelligent systems able to simulate many human traits. With this new era of simulated and artificial intelligence, much research is needed in order to continue to advance the field and also to evaluate the ethical and societal concerns of the existence of artificial life and machine learning.

The **Advances in Computational Intelligence and Robotics (ACIR) Book Series** encourages scholarly discourse on all topics pertaining to evolutionary computing, artificial life, computational intelligence, machine learning, and robotics. ACIR presents the latest research being conducted on diverse topics in intelligence technologies with the goal of advancing knowledge and applications in this rapidly evolving field.

COVERAGE

- Computer Vision
- Intelligent Control
- Synthetic Emotions
- Algorithmic Learning
- Evolutionary Computing
- Neural Networks
- Agent technologies
- Machine Learning
- Artificial Life
- Cognitive Informatics

IGI Global is currently accepting manuscripts for publication within this series. To submit a proposal for a volume in this series, please contact our Acquisition Editors at Acquisitions@igi-global.com or visit: http://www.igi-global.com/publish/.

The Advances in Computational Intelligence and Robotics (ACIR) Book Series (ISSN 2327-0411) is published by IGI Global, 701 E. Chocolate Avenue, Hershey, PA 17033-1240, USA, www.igi-global.com. This series is composed of titles available for purchase individually; each title is edited to be contextually exclusive from any other title within the series. For pricing and ordering information please visit http://www.igi-global.com/book-series/advances-computational-intelligence-robotics/73674. Postmaster: Send all address changes to above address.

Titles in this Series

For a list of additional titles in this series, please visit:
www.igi-global.com/book-series/advances-computational-intelligence-robotics/73674

Principles and Applications of Adaptive Artificial Intelligence
Zhihan Lv (Uppsala University, Sweden)
Engineering Science Reference • copyright 2024 • 300pp • H/C (ISBN: 9798369302309) • US $325.00 (our price)

AI and Blockchain Applications in Industrial Robotics
Rajashekhar C. Biradar (Reva University, India) Geetha D. (Reva University, India) Nikhath Tabassum (Reva University, India) Nayana Hegde (Reva University, India) and Mihai Lazarescu (Politecnico di Torino, Italy)
Engineering Science Reference • copyright 2024 • 414pp • H/C (ISBN: 9798369306598) • US $300.00 (our price)

Emerging Advancements in AI and Big Data Technologies in Business and Society
Jingyuan Zhao (University of Toronto, Canada) Joseph Richards (California State University, Sacramento, USA) and V. Vinoth Kumar (Vellore Institute of Technology, India)
Engineering Science Reference • copyright 2024 • 320pp • H/C (ISBN: 9798369306833) • US $270.00 (our price)

Advanced Applications of Generative AI and Natural Language Processing Models
Ahmed J. Obaid (University of Kufa, Iraq) Bharat Bhushan (School of Engineering and Technology, Sharda University, India) Muthmainnah S. (Universitas Al Asyariah Mandar, Indonesia) and S. Suman Rajest (Dhaanish Ahmed College of Engineering, India)
Engineering Science Reference • copyright 2024 • 481pp • H/C (ISBN: 9798369305027) • US $270.00 (our price)

Artificial Intelligence in the Age of Nanotechnology
Wassim Jaber (ESPCI Paris - PSL, France)
Engineering Science Reference • copyright 2024 • 299pp • H/C (ISBN: 9798369303689) • US $300.00 (our price)

For an entire list of titles in this series, please visit:
www.igi-global.com/book-series/advances-computational-intelligence-robotics/73674

IGI Global
PUBLISHER of TIMELY KNOWLEDGE

701 East Chocolate Avenue, Hershey, PA 17033, USA
Tel: 717-533-8845 x100 • Fax: 717-533-8661
E-Mail: cust@igi-global.com • www.igi-global.com

Table of Contents

Detailed Table of Contents

any growing country. Initially, the farmers were using manual methods to regulate irrigation and fertilizers. Manual irrigation and fertilization is also a major reason for the huge loss of productivity. The irregularity of these two is the reason for major problems in an agricultural crop. The addressed issues can be resolved with some automated methods based on artificial intelligence. The proper implementation of improvements in irrigation and fertilization is crucial for the efficient utilization of these resources and also contributes to Sustainable Development Goals (SDGs). Consequently, a data-driven smart farming system has been proposed to resolve issues related to irrigation and fertilization. Here a hybrid deep learning technique has been employed in the system to provide prediction data.

Artificial intelligence in things (AIoT) has revolutionized the capabilities and features of smart gadgets. Technology advancements in smart sensing have allowed for the seamless integration of AI and IoT, increasing the general intelligence and efficacy of smart devices. This chapter looks at smart sensing technology's progress and challenges in the context of AIoT integration. The study begins with a brief introduction of AIoT and its significance in the smart device industry. It then delves into the numerous smart sensing technologies that aid in bringing AI and IoT together, such as environmental sensors, motion sensors, biometric sensors, and more. Miniaturization, improved accuracy, and lower power consumption are just a few of the ways that these sensor technologies have progressed. The study also highlights the challenges of integrating AIoT with smart sensing technology. The need for efficient power management, interoperability, the complexity of sensor fusion and data integration, and concerns over data security and privacy are just some of the obstacles in the way.

Due to its many attributes, including decentralization, transparency, and security, blockchain is now a very important and novel technology for the current generation of industrial applications. One of these fields, the cloud of things, was created by fusing cloud computing and the internet of things. Blockchain is a key component of this technology because it provides data privacy and network security, eliminates cloud of things challenges through decentralization, and incorporates cloud of things features through elasticity and scalability functionalities, enhancing the overall efficiency of blockchain operations. As a result, it is highly advantageous to connect blockchain to the internet of things (IoT); this technology is known as BCoT, and it is now thought to be a very promising field for a wide range of industrial applications.

Wireless sensor devices and IoT (internet of things) devices are two interconnected technologies that have revolutionised the way we interact with our surroundings. These devices are at the forefront of the digital transformation and are instrumental in creating smart and connected environments. This chapter first discussed the basics of wireless sensor devices and IoT devices, with pros and cons. Second, it discussed the major differences between the two; third, the major real-time applications of both types of devices are discussed, and fourth, the discussion about the use of AI and ML algorithms in these devices with detailed explanations of major algorithms are presented. Finally, there is a discussion about the future scope of both wireless sensor devices and IoT devices with AI and ML. Through this chapter, one can learn about different kinds of devices for different applications with various AI and ML algorithms. The future smart world requires many more smart devices for various applications. This chapter can provide a clear understanding of various devices and applications for the future smart world.

This chapter gives an insight view of the evolving technologies related to the connected home experience using different enabling technologies such as RFID technology as an application of internet of things (IoT) for digitally connected-home facilities. IoT (internet of things), 5G, and the smart wireless network are pushing the real-world interface close to the digital world. Although the gateway of such interface is going to act as a DAC/ADC that is available in electronics designs, such developments are going to bring a better experience of the human-machine establishment by enabling the various technology towards the virtual world with reliable and seamless

Preface

In the fast-paced world of technology, the fusion of Artificial Intelligence (AI), Internet of Things (IoT), and smart sensing technologies is reshaping the way we interact with our environment and devices. The advent of pervasive sensing, machine learning, cloud computing, and energy-efficient solutions has unlocked a new era of possibilities beyond the conventional realm of the mobile Internet. These advancements are ushering in exciting developments in areas such as smart cities, sustainable systems, IoT-oriented wellbeing applications, and more. At the same time, they introduce new challenges in terms of reliability, quality of service, and energy efficiency, which captivate the attention of researchers and professionals alike.

This edited reference book, *AIoT and Smart Sensing Technologies for Smart Devices*, serves as a nexus for scholars, researchers, and practitioners to converge, exchange ideas, and share their insights on the remarkable progress in smart technologies. It navigates the terrain of advanced communication systems, novel application areas, and the burgeoning ecosystem of new devices and smart things that are increasingly connecting to the Internet through a myriad of wireless technologies.

In an era characterized by the rapid evolution of technology, it is imperative to foster collaboration and knowledge exchange among experts from diverse backgrounds. This book is designed to facilitate this exchange and provide a platform for exploring the latest innovations and best practices in the world of AIoT and smart sensing technologies.

ORGANIZATION OF THE BOOK

In Chapter 1, we delve into the transformative potential of AI and blockchain technology in the business world. As companies evolve, AI offers a unique opportunity to enhance various processes, driving innovation and improving customer and employee interactions. We explore the concept of Business Innovation based on Artificial Intelligence and Blockchain Technology (BI-AIBT) as a means to boost operational efficiency and maintain secure client relationships. By presenting qualitative empirical data from two different business sectors, this chapter highlights the promising synergy between AI and blockchain for the future of business.

Chapter 2 delves into the agricultural industry is a cornerstone of any nation's economy, and advancements in automation are crucial for its growth. This chapter examines how AI can revolutionize agriculture by addressing key issues like irrigation and fertilization. A data-driven smart farming system is proposed, integrating hybrid deep learning techniques to provide predictive data. By improving resource utilization and contributing to Sustainable Development Goals (SDGs), AI-driven smart farming has the potential to transform agriculture and enhance its sustainability.

In Chapter 3, AIoT has revolutionized smart devices by seamlessly integrating AI and IoT. This chapter explores the progress and challenges of smart sensing technology in the context of AIoT integration. It discusses various smart sensing technologies, their miniaturization, accuracy, and power consumption improvements. However, it also addresses challenges such as power management, interoperability, data integration, and privacy concerns. This chapter provides a comprehensive overview of the evolving world of AIoT and smart sensing technologies.

In Chapter 4, we talk about blockchain technology is a crucial component of the Cloud of Things (CoT), which combines cloud computing and IoT. This chapter explores how blockchain enhances data privacy, network security, and overall efficiency in CoT applications. By eliminating challenges and incorporating CoT features, blockchain technology has the potential to create a more secure and scalable environment. The chapter introduces the concept of BCoT, offering promising opportunities for industrial applications.

Chapter 5 delves into wireless sensor devices and IoT have transformed our interaction with the environment. This chapter provides a fundamental understanding of these technologies, highlighting their real-time applications and the integration of AI and machine learning algorithms. It also discusses the future prospects of these technologies in creating a smarter world with various devices and applications.

Chapter 6 talks about the connected-home experience is evolving with technologies like RFID and IoT. This chapter offers insights into how these enabling technologies are shaping digitally connected homes. It explores the interface between the physical and digital worlds, the role of IoT, 5G, and smart wireless networks, and their impact on various sectors, including hospitality, retail, automotive, and health.

Chapter 7 delves into building on the earlier exploration of BCoT, this chapter delves deeper into how blockchain is an integral part of the Cloud of Things (CoT). It further emphasizes the advantages of BCoT in a wide range of industrial applications. The chapter provides a comprehensive understanding of the symbiotic relationship between blockchain and IoT within the CoT framework.

Chapter 8 discusses early detection of Parkinson's disease is vital for effective treatment. This chapter explores the use of deep learning and machine learning in detecting PD based on speech signals. A comparative analysis of different models, including CNNs, LSTMs, Random Forest, and XGBoost, is presented, highlighting their performance in terms of accuracy, precision, recall, and F1-score.

Chapter 9 discusses a world where time and energy are precious. This chapter focuses on automating vehicle parking systems with minimal human interference. It introduces a data-driven smart parking system that utilizes IoT for real-time parking slot availability information. The system provides an innovative solution to optimize parking space usage and enhance the overall parking experience.

In Chapter 10 routing protocols play a crucial role in wireless sensor networks and IoT applications. This chapter provides a systematic analysis of existing routing protocols, addressing the choices and challenges associated with data transmission efficiency and reliability. It covers various routing methods, from flat routing to hierarchical routing and energy-aware routing, highlighting their applications and needs. The chapter also discusses performance optimization, clustering protocols, and open research issues in the field.

These chapters collectively offer a comprehensive exploration of AIoT and smart sensing technologies, showcasing their potential to revolutionize various domains and addressing the challenges they present. We invite you to delve into the wealth of knowledge contained within these chapters and embrace the exciting possibilities of this rapidly evolving field. Our contributors span a spectrum of expertise, addressing a broad array of topics that span from IoT-oriented wellbeing applications to wearable communications, from smart metaverse apps to the sustainable development goals of our planet. We have gathered research contributions that tackle challenges in IoT-oriented medical applications, routing in wired/wireless environments, energy harvesting in smart cities, and so much more.

The intended audience for this book includes experts, researchers, and students who are passionate about AIoT and smart technologies and their profound impact on the way we live, work, and interact with the world. We have also compiled a list of key topics that represent the diverse range of subjects covered within the book, ensuring that it caters to the interests of a broad readership.

As editors of this comprehensive volume, we are honored to present these cutting-edge contributions that explore the ever-expanding frontiers of AIoT and smart sensing technologies. We invite you to delve into the wealth of knowledge contained within these chapters, and we believe that this book will serve as a valuable resource for anyone seeking to stay at the forefront of technological advancements and their transformative potential.

Sincerely,

Fadi Al-Turjman
AI and Robotics Institute, Near East University, Nicosia, Turkey & Faculty of Engineering, University of Kyrenia, Kyrenia, Turkey

Chapter 1
A Description of How AI and Blockchain Technology Are Used in Business

Ramiz Salama
Near East University, Nicosia, Turkey

Fadi Al-Turjman
AI and Robotics Institute, Near East University, Nicosia, Turkey & Faculty of Engineering, University of Kyrenia, Kyrenia, Turkey

ABSTRACT

The ability to build new kinds of cooperation (AI) is one way that the numerous company processes may be improved as a result of the growing firm evolution and the most recent artificial intelligence. Such developing technologies enable the provision of brand services and even some cutting-edge forms of customer and employee interactions in the workplace. Businesses were simultaneously forced to concentrate on their current strategies and actively seek out new market chances on a regular basis as a result of AI digitalization. Blockchain technology can protect data privacy, even though digital technology research in the context of corporate innovation is receiving more attention. This study presents BI-AIBT, or business innovation based on artificial intelligence and blockchain technology, with the aim of increasing corporate operations and upholding secure client relationships. The qualitative empirical data collection consists of a small number of major respondents from two different business sectors.

DOI: 10.4018/979-8-3693-0786-1.ch001

INTRODUCTION

Artificial intelligence (AI) is a field of research that aims to create machines that are intelligent enough to do complicated activities without the need for human intelligence (Rahman, 2022). As it encourages the adoption of cutting-edge technologies in the Fourth Industrial Revolution (IR 4.0), such as blockchain, cryptocurrency, cloud computing, and the internet of things (IoT) among others, AI is one of the driving forces behind industrial development. In reality, the proliferation of IoT devices, social media, and online apps has sped up the advancement of AI. This data is employed to train machine learning algorithms. However, there are several issues with AI. Particularly in light of countless data breaches and other cases of abuse, privacy has grown to be a serious issue. The Facebook incident, in which the political consulting company Cambridge Analytica unintentionally targeted millions of users, is one instance of this. The technology cannot be evaluated or trusted since it is unable to interact with or communicate with human users—another concern with AI that is becoming more and more prevalent (Cheng, 2022).

Similar to this, blockchain is growing in popularity as a technology with numerous potential uses in a variety of industries. Blockchain, a revolutionary concept that gained notoriety with the creation of bitcoin in 2008, is still transforming a variety of aspects of our lives, including how we interact, keep track of transactions, and make payments automatically. In reality, the blockchain has enabled its users to send money directly among themselves in a safe and secure manner. Additionally, because blockchain is a shared database that is synced across various locations, the execution of smart contracts may make it easier to check rights and compliances. Using consensus techniques like proof of stake or proof of work, blockchains store data in blocks with hash values and timestamps. Proof of stake is less expensive than proof of labor due to its superior energy efficiency. However, the ability for all mining nodes that have a copy of the whole ledger to independently validate each transaction is made feasible by the fact that every transaction on the blockchain is cryptographically signed. As a result of not requiring a centralized authority to authenticate transactions, blockchain is secure and cost-effective.

The acceleration of AI and blockchain integration will change the IR 4.0-inspired future digital generation. Blockchain-based solutions, can provide explainability, privacy, and trust to AI-based applications, whereas AI can improve scalability and security while addressing customization and governance issues. Even though blockchain and AI are conceptually distinct in a number of ways, they can be combined to make up for each other's weaknesses. This is how AI and blockchain function as the yin and yang of digital trade, with blockchain facilitating execution, verification, and recording and AI supporting comprehension, recognizing, and decision-making. As the term suggests, big data refers to enormous amounts of data that are difficult to manage using regular software or web-based solutions. Google will have processed 40,000 queries per second and 44 trillion gigabytes (GB) of data. 3.46 million Searches are performed every day, for a yearly total of 1.2 trillion. The capacity for storage, processing, and analysis has all been expanded. There are numerous definitions of big data, but the most popular and well-known one is by Douglas Laney. Laney emphasized the growth of (big) data in terms of the three "V's" of volume, velocity, and variety. Big data's "huge" component refers to its volume. Patients can define the conditions under which certain researchers may temporarily access specific portions of their medical records with the aid of blockchain technology. Patients can connect to other organizations and have their medical information automatically collected using blockchain technology. Because of how transparent it is, information may be sent in a secure manner. Clinical trial data, monitoring equipment, electronic medical records for sharing and keeping, mobile health apps, and cloud storage for insurance data are all examples of blockchain applications in the healthcare industry. Blockchain is stable because it would take an unimaginably large amount of processing effort to refigure the altered block and all future blocks if someone wanted to change a block after it had been added to the chain. Along with volume, big data also relates to velocity and diversity. The rate or pace at which data is gathered and made accessible for additional processing is known as velocity. Every business or system has the ability to collect transaction-level data in addition to other types of structured and unstructured information like audio, video, text, or log files. The industry now uses these three "Vs" to describe big data. Big data has recently become incredibly popular all around the world. For a variety of reasons, nearly all study domains, both in academia and in business, produce and analyze enormous amounts of data. The hardest challenge is managing this enormous amount of data, which may be both structured and chaotic (Kharbat, 2020).

With such a huge amount of data, using artificial intelligence (AI) algorithms and cutting-edge fusion techniques would make sense. Automating decision-making using machine learning (ML) techniques like neural networks and other AI algorithms would be a significant advancement. Instead of directly coding computer programs using algorithms and statistical models, machine learning research focuses on developing computer systems that learn through inference and patterns. Machine learning has advanced significantly during the past ten years. The most crucial components of machine learning are the data, which form the basis of any model. The forecasts become more accurate as the data becomes more relevant. To get more precise forecasts, we must select an algorithm based on the information and the situation. Blockchain is secure because several copies of the data are kept on different nodes, eliminating the chance of an adversarial user assaulting a centralized structure. These characteristics make blockchain useful for handling healthcare data. It is regarded as a promising method of securely transmitting medical data. However, not every issue involving extremely sensitive data can be solved by blockchain. It has been questioned if blockchain technology has a place in the healthcare industry. Governments may wish to look into potential blockchain applications in the healthcare sector given the technology's recent debut and take into account the challenges posed by the industry's traditionalism. A thorough analysis of the trade-offs is required when developing and deploying blockchain solutions for the healthcare industry (Rahman, 2022). We aim to discuss how machine learning, big data, and blockchain technology operate in this environment as well as potential applications in the fields of medicine, healthcare, and public health.

EXTENT OF PAST WORKS

The current study makes use of bibliometric analysis to assess the intellectual organization and performance (such as productivity of publications, most well-known articles) of the literature on AI and blockchain integration in business. In order to comprehend a body of literature, bibliometric analysis essentially uses quantitative approaches to analyze bibliometric data, such as publication and citation statistics. The approach is a tried-and-true technique for doing scientific research that can be used to a variety of fields, including business. There are numerous methods for reviewing the literature, but the bibliometric method relies on quantitative analytical methods and a review protocol, making it the most objective. Other review methods, such as critical, either lack review protocols or are only capable of providing subjective interpretations because they lack objective analysis methods, such as thematic analysis. The current study explicitly adheres to the four-step bibliometric analysis methodology, which comprises deciding on an analysis strategy, choosing the right data to use, conducting the analysis, and reporting the findings. The study's methodology is outlined in its broadest sense. The current study combines bibliometric and content analytic techniques to follow the evolution of the business-related literature on AI and blockchain integration (Qasim & Kharbat, 2020). Blockchain technology, business data analytics, and artificial intelligence: Use in the accounting profession and ideas for inclusion into the accounting curriculum. Journal of emerging technologies in accounting. In order to pinpoint the major themes and topics in the research field, the study specifically conducts a performance analysis using a number of bibliometric indicators, such as citations, cite score, impact factor, publication count, and h-index. After this research, the authors read the full text again to conduct a content analysis.

Business innovation occurs when a company implements novel practices, products, or services that profitably expand a market. Business innovation is the process by which a company introduces new trends, services, or other products to foster firm growth. It can entail kicking off the project without any previous planning and utilizing tried-and-true techniques. Technology that has been trained to perform intelligent tasks that have historically been created by people is referred to as AI. Blockchain is a decentralized computer network that collects and stores data to display an event's chronological evolution on a simple and immutable record basis. This innovation can need either altering current practices or developing a whole new method. Blockchain, for instance, may boost confidence, accountability, protection, and privacy in business units by offering a shared and decentralized distributed directory. A blockchain is similar to a register or well-known distributed ledger in that it may store a wide variety of features. A distributed ledger is a set of data that is freely exchanged and synchronized by multiple people over numerous locations, businesses, or topographies. The same technology that is used by blockchain and bitcoin is known as basic appropriated records. A hash, a type of irreversible cryptographic stamp, is used in the blockchain DLT system to track transactions. Every each square in the subsequent block-based compilation of the transactions has a hash of the square before it, linking them together. This leads to the word "blockchain" being occasionally used to refer to transmitted records. These data have a strong connection to both identities and money. The IoT is developing in German and European businesses to enhance industries and business processes. Not to mention, AI improves operations by recognizing and enhancing the results of business processes. The goal of the innovation process is to increase a company's bottom line. By creating new sales opportunities, increasing revenue on existing platforms, saving time and resources, and improving output or performance.

Because existing expertise would become outdated owing to technology advancement, the blockchain invention eliminates it. All business structures in all industries would change if this revolutionary architectural breakthrough were to be widely adopted. Blockchain enables the utilization of processing power, data, and algorithms for various artificial intelligence components, including decentralized markets and collaborative platforms (Kumar et al., 2023). It might encourage the usage of artificial intelligence (AI) and several other technical developments. Artificial intelligence (AI) refers to tools created in the past to do mental tasks. A decentralized computer network known as blockchain records and saves data in an immutable, transparent ledger system that shows an event chronology. Combining AI and blockchain can enhance machine learning and give AI access to financial resources. Among other things, the blockchain enables safe data exchange and storage. Manufacturing supply chains could undergo a significant transformation as a result of the development of blockchain technology. Middlemen could be eliminated, operations could be reorganized, overall security would increase, and information governance would improve. Artificial intelligence (AI) has the potential to reduce time and costs at work by streamlining and automating repetitive tasks and procedures. A rise in efficiency and productivity overall enables quicker management decisions based on cognitive technology results. Businesses might benefit from using artificial intelligence to provide more individualized customer experiences. Big data analysis is substantially more effective when using AI. It can quickly identify trends in the data, including previous purchases, preferences, credit scores, and other pertinent elements. Building a solid brand and marketing the ideas of digitization and artificial intelligence are challenging tasks. AI is widely employed in marketing initiatives where speed is essential. Without the assistance of marketing professionals, AI systems might be able to engage with customers more successfully based on information and client data, then deliver personalized messaging at the right time to achieve maximum productivity. Artificial intelligence can help businesses better manage their online reputation and protect their current online presence. Giving brand and service executives access to websites, social networking sites, and other locations allows them to undertake in-depth assessments and research. Conducting the bibliometric-content analysis and presenting the results are the final steps. The authors do bibliometric analysis and network the results using the programs Gephi and VOSviewer. To enhance the presentation of their data, the authors also employ a Python script to generate word clouds that capture the key ideas in bibliographic clusters. Jupyter Notebook and Python 3.7 were both used to run the script and create the word clouds. Each article in each significant subject cluster identified by the bibliometric study was read in its entirety and fairly evaluated. The sections that follow present the results of the bibliometric–content analysis.

HOW AI AND BLOCKCHAIN TECHNOLOGY ARE USED IN BUSINESS

AI (Artificial Intelligence) and blockchain technology are two separate but powerful technologies that can be used in various ways to enhance and transform businesses. Here's an overview of how they are employed in business:

AI (Artificial Intelligence) in Business

Customer Service and Support: AI-powered chatbots and virtual assistants can provide 24/7 customer support, answer frequently asked questions, and handle routine inquiries, improving customer satisfaction and reducing response time. Data Analysis and Insights: AI can analyze large datasets to identify trends, patterns, and correlations that may not be apparent to human analysts. This helps in making data-driven decisions and improving strategies (Loureiro et al., 2021).

Personalization: AI algorithms can analyze user behavior and preferences to provide personalized recommendations, content, and product suggestions, enhancing the customer experience.

Automation: AI can automate repetitive and time-consuming tasks, such as data entry, processing invoices, and managing inventory, leading to increased efficiency and cost savings.

Predictive Maintenance: In manufacturing and logistics, AI can predict when equipment or machinery is likely to fail, enabling proactive maintenance and reducing downtime.

Fraud Detection: AI can identify unusual patterns and anomalies in financial transactions, helping to detect and prevent fraud.

Natural Language Processing (NLP): NLP is used for sentiment analysis, content summarization, language translation, and more, making it easier to understand and engage with customers and stakeholders.

Healthcare and Life Sciences: AI is used for drug discovery, disease diagnosis, patient monitoring, and optimizing healthcare operations.

Marketing and Advertising: AI helps in targeted advertising, optimizing ad campaigns, and analyzing consumer behavior to improve marketing strategies.

Blockchain Technology in Business

Supply Chain Management: Blockchain can be used to create transparent and tamper-proof supply chains, allowing businesses and consumers to track the origin and journey of products from manufacturer to end user.

Smart Contracts: Smart contracts are self-executing contracts with the terms of the agreement between buyer and seller directly written into code. They automate contract execution and reduce the need for intermediaries.

Digital Identity Verification: Blockchain can be used for secure and decentralized digital identity verification, making it easier to establish trust in online interactions and reducing identity theft.

Cross-Border Payments: Blockchain can facilitate faster and more cost-effective cross-border transactions by eliminating intermediaries and reducing currency conversion fees.

Intellectual Property Protection: Businesses can use blockchain to prove the authenticity and ownership of intellectual property, including patents, copyrights, and trademarks.

Voting Systems: Some countries are exploring the use of blockchain for secure and transparent voting systems, reducing the risk of fraud and ensuring the integrity of elections.

Real Estate and Property Records: Blockchain can streamline the process of property transfer, record-keeping, and title management by ensuring a secure and immutable ledger.

Tokenization of Assets: Businesses can tokenize assets like real estate, art, or company shares on a blockchain, making them more accessible and liquid for investors.

Data Security and Privacy: Blockchain can enhance data security and privacy, allowing individuals to have greater control over their personal information.

Both AI and blockchain technology offer a range of opportunities for businesses to improve operations, enhance security, and create new revenue streams. The choice to implement them depends on specific business needs, goals, and the industry in which the business operates.

RESULTS AND DISCUSSION

The social and economic institutions that support businesses as well as the way they are operated have the potential to be significantly changed by blockchain technology. We'll look at how, in some circumstances, block chain can be utilized to quickly validate a transaction's properties. The blockchain provides an authenticating system that is reliable. Your customers' and your employees' electronic IDs make it simple to verify their identities. When this data is kept on a public blockchain, the risk of identity fraud, financial fraud, theft, and other cybercrimes is decreased. Blockchain and artificial intelligence-based business innovation (BI-AIBT) have been proposed as ways to improve corporate operations and provide a secure link with a variety of clients. In the collection of qualitative analytical data, there are just a few noteworthy respondents from two different business sectors. The corporation evaluated BI-AIBT and contrasted and compared how digitization affected value generation, proposal, and business capture. Additionally, BT can improve the way organizational strengths and talents interact. A thorough analysis of the literature and localization strategies for peer-reviewed studies serve to highlight the business innovation. SMEs can experience sustainable productivity development in the modern economy if new technologies are employed in the workplace. By implementing a number of strategies, including working with other businesses and academic organizations, businesses can become even more imaginative. Additionally, the majority of SMEs in Moldova are unaware of the need to market research findings and implement innovations based on scientific discoveries. Big-data analytics (BDA), should be given much more attention since it may support business planning and provide useful data for boosting service innovation. They investigated the application of thorough service innovation analytics. Actual customer requests were gathered through the digital service channel by the emphasis using analytical data characteristics. Research on business transformations, applied mathematics, business engineering, financial science, and the global business infrastructure is done using the applied holistic mathematical model (AHMM) developed by Trad et al. Chandrachai, Triukose, One instance of how the suggested formality of AHMM replicated many aspects of human cognition was the heuristic decision-making process, which mostly depended on beam examination. By supporting a thorough separation process, synchronization of various EA methodologies, and transition strategies, the AHMM facilitated business transformation operations. It is unique because of this.

The model-narrative review (MNR) technique, according to Hakala et al, may link extensive work on the market, entrepreneurial, and innovation contexts to comprehend this evolving narrative. They contributed to the opening up and conceptual clarity of the language around the environment by disclosing a number of implicit interpretations and basic presumptions. MNR as a whole synthesizes and carefully considers the parallels and differences among linked ideas, resulting in the development of a method for developing model narratives that offer room for alternative research studies.

Improvised marketing interventions (IMIs) are defined as social media actions that are created and executed in real-time close to an actual incident. Five multi-method assessments were conducted; their findings were recorded along with analytical information and simulated tests. The IMI's effectiveness was evidenced by the laughter and unexpected results. The findings showed both the IMI's potential for social media and the attributes that companies need to have in order to benefit from proactive online networking. According to Mustafa et al. Typical FinTech companies may present opportunities as well as challenges for Islamic banking systems. According to the results of this investigation, Islamic financial institutions are more likely to see Islamic FinTech firms as partners than as rivals. Adaptive Hybridized Intelligent Computational Models (AHICM) were developed by to study consumer behavior for business expansion. A detailed examination of market segmentation and customer needs is necessary for product advancements, innovative concepts, and societal revolutions. The blockchain-based integrative safety mechanism (BISM), which intends to provide secure access management and privacy protection for both products and individuals, is described by Manogaran et al. (2020). While privacy protection depends on the robustness of response times, user access procedures are based on the state of virtual resources at different points in time. SMEs, BDA, AHMM, MNR, and IMI are among the established business strategies that have been outperformed by Business Innovation based on Artificial Intelligence and Blockchain Technology (BI-AIBT). The demand forecast ratio, product quality ratio, business development ratio, customer behavior analysis ratio, and customer satisfaction ratio all have potential for improvement, according to BI-

AIBT. A data-collection technique called blockchain makes it hard or impossible to alter, hack, or cheat the system. A blockchain is a collection of duplicate digital ledgers that have been dispersed over the whole computer network. Blockchain, also known as Distributed Ledger Technology (DLT), uses decentralization and cryptographic hashing to make it simple and impossible to change the past of a relatively complex resource. The disclosed method concurrently grants everyone access to the record by creating a decentralized appropriation chain. Only a few non-financial uses for blockchain technology include supply chain management and digital identity. Supply chain management promotes effectiveness, teamwork, and superior outcomes. Additionally, it lowers costs. It encourages companies to better meet customer demands and more swiftly manage demand, stock distribution, interference management, and expense control. With the use of blockchain technology, consumers may have more control over their own actions. There is no chance for any focus point to take the client's personality into consideration, and customers must consent before a company may use their information. The most current research demonstrates the benefits of combining blockchain technology with IoT and AI. For instance, the system architecture of many Internet of Things (IoT) devices could be improved by the use of blockchain technology. Making judgments using computers IoT creates flexible instruments that effectively mimic smart leadership with almost no human resistance. While IoT manages their internet communication, AI allows the devices to learn from their data and experience. IoT makes it possible for Internet-connected devices to communicate information with private blockchain firms so that secure records of shared transactions may be updated. Thanks to IBM Blockchain, your coworkers can access and share IoT data with you, but only you and the leaders will need to serve as the primary central authority. Most of the time, the interrelationships between blockchain, IoT, and AI are neglected, and each technology is applied on its own. These technologies could, and in a perfect world, ought to, be developed simultaneously. There may be a connection between IoT and AI in that while IoT gathers and feeds data, AI improves corporate processes and legal frameworks through Blockchain, constructing infrastructure and defining engagement standards. It demonstrates how blockchain technology and artificial intelligence are being applied in business. Coordination of commercial transactions between demand and resource providers may be accomplished using blockchain technology. The resources that are available will determine the project manager who will put together the tools required to do the assignment. Finding stronger company motivators is the main benefit of having solid, healthy supplier relationships. The likelihood that you will acquire enthusiastic dedication, correct evaluation, and excellent terms increases as you and your suppliers become more familiar with one another. In this scenario, blockchain is primarily utilized to sustain supply chain connections between the needs of many consumers and many providers. Customers can create new demand using the decentralized order engine, and producers can provide new items as part of a new transaction using the decentralized product engine. Because a blockchain's centralization or decentralization often pertains to the rights of participants in the record, it is also a design issue. Anyone can participate and make decisions in a decentralized organization. Only recognized and well-known groups are allowed to take measures on the record in a united organization. To confirm and secure the data patterns for transactions between providers and consumers, the smart contract mechanism serves as an authenticating protocol between the two engines. Commercially accessible technology might not always be able to offer the necessary level of security. "Commercial-off-the-rack" (COTS) programming is a term used in the business world to describe programming that is immediately accessible for purchase. In order to upgrade, replace constraining frameworks, and carry out other duties, public area associations are increasingly relying on COTS applications. A protection plan that is mission-based is a solution that can be adjusted to meet specific needs. It serves as an example of how an additional layer of specialized security could safeguard the data on a company's network. This personalized method has the advantage of allowing the employee's agent (EA) to control the rate of public and private key renewal and monitor key creation parameters. Employee agents (EAs) are persons who have been given permission to act as another person's agent by their employer. For instance, even when no other representatives are permitted, you can designate a worker as an authorized expert to allow them to make substantial purchases on the company's behalf.

Accounting for indirect costs is known as ICA. The ICA only grants authorized EAs restricted access to each database cluster. a method of gathering complex data that needs at least one mathematical operation and a secret key or other "key" to decode the data To prevent identity theft and extortion, websites that communicate sensitive data, such as bank account numbers and Mastercard numbers, should always encrypt the data. Before entering the database, the data is encrypted. Each data cluster is encrypted with a different key in order to restrict access to the data to only authorized workstations. Thanks to ICA, the firewall in the receiving state can access customer and business data from the internet or extranets. The secret key to utilize when symmetrically encrypting data is specified by the ICA. Secret keys are generated for symmetric key encryption while the ICA is processing in this mode. Information about corporate and customer accounts can be stored in databases using passwords. The primary delivery's personnel are under the ICA's control. Location or account balance are the two ways that EAs can be assigned to an account (Manogaran et al., 2020). While in the transmitting state, the ICA sends the secret keys of the associated EAs. Using a secret key, the ICA encrypts data collected from external networks. The ICA sends data to the clustered database. The company accounts for each employee's workstation are under their control. To exchange data with other team members, the sender encrypts it using the recipient's public key and decrypts it using the recipient's private key. Data is protected from internal illicit use by this upgraded authentication layer. It also demonstrates how blockchain marketing has evolved. Its key focuses are peer-to-peer communication, shifting market dynamics, and getting rid of middlemen that limit and filter data sources and raise costs. Intermediaries are helped by this mediation. By facilitating access to data and generating widely used, immutable data archives, blockchain technology can enhance the quality of data. Information quality includes a wide range of components that affect an association's ability to function. Reports are unreliable when data isn't acquired properly. Frameworks with divergent preset values obstruct real risk awareness. Clarvonyt aids in resolving these problems and improves each piece of work. Blockchain technology has the ability to drastically impact corporate experiences through data exchange, information privacy, and a customer-focused strategy. Innovative plans are put in place to ensure client pleasure and potential value. With the development of the internet, business practices for marketing their goods and services have changed. Due to technology, which has also produced new electronic intermediaries, traditional middlemen have lost some of their significance. Access to the internet has also made it easier for new online intermediaries to provide customers innovative goods and services Sheron, Sridhar, Baskar & Shakeel, 2019). In order to respond to RQ1, which asks what the publishing productivity of research on AI is and blockchain integration for business, the study examines all of the publications in the field organized by year. Scopus provided the bibliometric data that was used in the inquiry. Information on scientific publications, including publication details (such as the title, abstract, keywords, and year) and citation details (such as the author, document, and journal citation total) are included in a subset of big data called bibliometric data (Broadus, 1987; Donthu et al., 2021).

CONCLUSION

Artificial intelligence (AI) and blockchain are two of the core IR 4.0 technologies. Despite the fact that the two technologies are quite different from one another, amalgamation is a topic that interests both academics and practitioners. However, there is a dearth of research that compiles and assesses the most recent findings regarding the commercial uses of blockchain-AI systems. To fill this gap, the current study used a bibliometric-content analysis and came up with five significant findings. The article makes it clear that AI and blockchain are IR 4.0 technologies that were initially referenced in 2016, with early research fusing the two technologies for commercial application first emerging in 2017 but only exploding from 2019 onwards. The conclusion of the study that the marriage of AI and blockchain offers substantial promise for commercial applications is supported by the wealth of research that has been seen in this subject. Second. The Liu et al. article on the application of AI and blockchain in healthcare. Work on a blockchain-based LSTM credit evaluation system's article on performance optimization for blockchain-enabled IoT systems. The primary works in the field should be familiar to prospective authors in order to position and arrange future research in unique ways to expand understanding of earlier study. Third, the study reveals that the three most popular research areas are "blockchain and machine learning," "blockchain and smart contracts," and "blockchain and security." The finding that "blockchain and machine learning" is the strongest association in the corpus and that machine learning is the AI technology that seems to be most pertinent and salient is another way to use triangulation to highlight the promise and significance of integrating blockchain and AI in business. Fourth, the paper breaks down four key topic groupings that serve as the conceptual foundation of research on the integration of AI and blockchain for business, including IR 4.0 and supply chains, smart healthcare, safe transactions, and finance and accounting. Three of the study's clusters—safe transactions, supply chains, and IR 4.0—are generic and cross-industry, while the other two (smart healthcare and finance and accounting) are separate and industry-specific. Because there aren't many large subject clusters, research on AI and blockchain integration for business is still in its early stages. Research that promotes the field's continuous evolution to improve current clusters and promote the emergence of new clusters is therefore highly welcomed. Fifth, the report identifies eleven businesses where prior research on the topic has advised combining blockchain with AI. These sectors include marketing, management, supply chain management, e-commerce, finance and accounting, healthcare, and intellectual property rights. In four separate ways, this post has made a substantial contribution toward that goal. By mapping the publication productivity of AI and blockchain integration for business, this study first demonstrates that the subject is still evolving and has plenty of room for expansion and more research. Second, by presenting the major pieces of research in the area, this article has opened the door for future studies to build on the groundbreaking findings presented here. Third, this page gives aspiring authors a current overview of the literature in the area by grouping the most popular issues and themes about AI and blockchain integration for business. Finally, our study has found 10 commercial possibilities for blockchain and AI integration. These applications should be helpful to business executives looking to use academic research to pinpoint potential opportunities for organizational transformation using the two IR 4.0 technologies. This essay has significant ramifications for a wide range of stakeholders, including corporate managers, AI and blockchain technologists, IT vendors, and aspiring academics. Blockchain and AI developers may obtain clearer understandings of how combining these two technologies improves corporate effectiveness. Blockchain and AI developers may work together to create automated, decentralized business apps that provide better client privacy and confidentiality protection, higher performance, and better governance. Additionally, by identifying the market for AI and blockchain business solutions and emphasizing the benefits they offer, IT providers will be better able to comprehend how to integrate AI and blockchain into enterprises Finally, by building on the current overview of AI and blockchain integration in business and exploring new topics that would generate new streams of study or enrich existing streams of research in the field, future academics will be able to enhance knowledge in the subject. The prospects for merging AI and blockchain to improve the expansion, resilience, and robustness of corporate processes are made clear to professionals and scholars in this essay, which provides cutting-edge insights. But like other pieces, this one is aware of its limitations. The primary limitation on the data in this article is the accuracy and completeness of the source, in this case Scopus. Remember that Scopus is a scientific database and that it may include (unintentional) errors since it was not intended for bibliometric research. To the best of their ability, the authors reduced any (unintended) errors by eliminating duplicates and erroneous entries from the bibliometric data they obtained from Scopus, as recommended. Second, the combination of blockchain technology and artificial intelligence represents a revolutionary advancement in the industry. There will likely be a large increase in its integration and application in this field, which will likely lead to the emergence of new study areas. As a result, prospective authors should use the provided search term in addition to the review insights provided here to stay current on new developments in the field.

The study presented here highlighted the dearth of insights emerging from business research, and future researchers on the business use of AI and blockchain integration are urged to focus their efforts on figuring out how such an integration might be applied from a business rather than an engineering lens. This article calls for additional study that addresses the non-exhaustive research challenges on AI and blockchain integration using a business lens generally and by cluster, as we highlighted, in order to support such initiatives.

REFERENCES

Akter, S., Michael, K., Uddin, M. R., McCarthy, G., & Rahman, M. (2022). Transforming business using digital innovations: The application of AI, blockchain, cloud and data analytics. *Annals of Operations Research*, *308*(1-2), 1–33. doi:10.1007/s10479-020-03620-w PMID:35935743

Attaran, M., & Gunasekaran, A. (2019). *Applications of blockchain technology in business: challenges and opportunities*. Academic Press.

Donthu, N., Kumar, S., Mukherjee, D., Pandey, N., & Lim, W. M. (2021). How to conduct a bibliometric analysis: An overview and guidelines. *Journal of Business Research, 133*, 285–296. doi:10.1016/j.jbusres.2021.04.070

Kumar, S., Lim, W. M., Sivarajah, U., & Kaur, J. (2023). Artificial intelligence and blockchain integration in business: Trends from a bibliometric-content analysis. *Information Systems Frontiers*, *25*(2), 871–896. PMID:35431617

Loey, M., Manogaran, G., & Khalifa, N. E. M. (2020). A deep transfer learning model with classical data augmentation and CGAN to detect COVID-19 from chest CT radiography digital images. *Neural Computing & Applications*, 1–13. doi:10.1007/s00521-020-05437-x PMID:33132536

Loureiro, S. M. C., Guerreiro, J., & Tussyadiah, I. (2021). Artificial intelligence in business: State of the art and future research agenda. *Journal of Business Research*, *129*, 911–926. doi:10.1016/j.jbusres.2020.11.001

Nirvan, S., Verma, S., Kathuria, S., Singh, R., & Akram, S. V. (2023, August). Enhanced Banking Services using Blockchain and Artificial Intelligence. In *2023 Second International Conference on Augmented Intelligence and Sustainable Systems (ICAISS)* (pp. 57-62). IEEE. 10.1109/ICAISS58487.2023.10250709

Qasim, A., & Kharbat, F. F. (2020). Blockchain technology, business data analytics, and artificial intelligence: Use in the accounting profession and ideas for inclusion into the accounting curriculum. *Journal of Emerging Technologies in Accounting*, *17*(1), 107–117. doi:10.2308/jeta-52649

Sheron, P. F., Sridhar, K. P., Baskar, S., & Shakeel, P. M. (2020). A decentralized scalable security framework for end-to-end authentication of future IoT communication. *Transactions on Emerging Telecommunications Technologies*, *31*(12), e3815. doi:10.1002/ett.3815

Wang, Z., Li, M., Lu, J., & Cheng, X. (2022). Business Innovation based on artificial intelligence and Blockchain technology. *Information Processing & Management*, *59*(1), 102759. doi:10.1016/j.ipm.2021.102759

Chapter 2
AIoT-Based Data-Driven Irrigation and Fertilization Smart System

Vinay Gautam
Chitkara University Institute of Engineering and Technology, Chitkara University, Punjab, India

Raj Gaurang Tiwari
https://orcid.org/0000-0003-1352-3321
Chitkara University Institute of Engineering and Technology, Chitkara University, Punjab, India

ABSTRACT

Agriculture is a prime industry and the foundation of a country's economy. Thus, advancement and automation in the agriculture industry is the foremost objective of any growing country. Initially, the farmers were using manual methods to regulate irrigation and fertilizers. Manual irrigation and fertilization is also a major reason for the huge loss of productivity. The irregularity of these two is the reason for major problems in an agricultural crop. The addressed issues can be resolved with some automated methods based on artificial intelligence. The proper implementation of improvements in irrigation and fertilization is crucial for the efficient utilization of these resources and also contributes to Sustainable Development Goals (SDGs). Consequently, a data-driven smart farming system has been proposed to resolve issues related to irrigation and fertilization. Here a hybrid deep learning technique has been employed in the system to provide prediction data.

DOI: 10.4018/979-8-3693-0786-1.ch002

INTRODUCTION

Agriculture in India is a huge industry and about 80% of the Indian population depends mainly on agriculture (MINAGRI, 2020). The finding of various studies is that the resources required for agriculture are limited and should be utilized properly without waste. Therefore, the farmer can afford to waste agricultural resources and low-quality agricultural produce. It must help them with such a solution that reduces their productivity loss and informs them to use resources properly. Several solutions have been proposed by researchers and scientists to help farmers. The maximum population of the farmer was not able to get benefits from them. The government has framed various policies such as ICT (Aker et al., 2016) for the farmer to provide information on time but the lack of communication is a major issue. ICT helps to provide valuable information to farmers to accelerate their productivity and enhance efficiency in work with various innovative methods. The farmer has benefited from ICT using various websites and other interfaces that provide valuable information to the farmer regarding the use of agricultural resources. With the help of ICT initiation, farmers can enhance their knowledge and abilities. After this initiative, the agriculture industry benefited from various solutions based on IoT and machine learning. The components of an IoT-based smart system are displayed in Figure 1.

IoT and machine learning are powerful technologies that are combined to enhance the capabilities and provide several solutions in data-driven farming (Kapitanova & Son, 2012). With the help of these two technologies, legacy agriculture practices have been replaced with smart farming (Mishra et al., 2019). The IoT and ICT have collaboratively been used to boost various systems used in agriculture. The collaborative solutions help farmers to optimize their valuable resources and to enhance their productivity in the agriculture industry. With the help of IoT infrastructure, the farmer can get valuable updates on climate and environmental factors. As IoT infrastructure, several devices and sensors have been deployed in the field to track various attributes of the environment and soil. The collected data from the farm field can be used for further study and transmitted through the Internet for immediate decision (Orn et al., 2020). The data collected from the farm can be analyzed from any place. The transmitted information is very informative for local farms to make a better decision. The data-driven Artificial Intelligence and IoT-based solution can be beneficial for the farmers and guide them daily. These solutions can help farmers provide a report of soil quality and guide them to optimize their valuable resources such as water and fertilizer. The data-driven AIoT-based solution for irrigation and fertilization for soil quality will be an innovative solution for farmers and resolve maximum issues related to the wastage of resources. This solution can help the farmer with real-time monitoring of irrigation and fertilization.

Figure 1. IoT-based smart agriculture

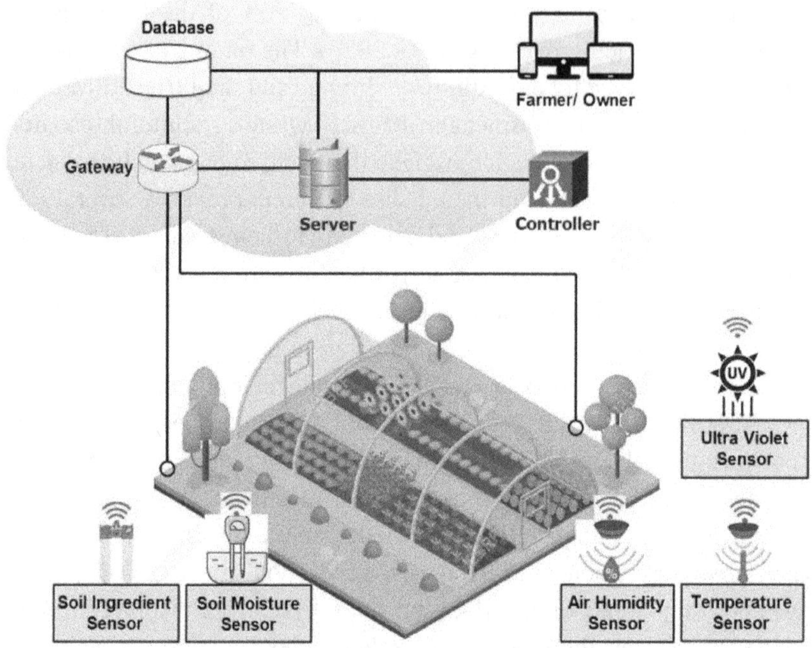

Artificial intelligence further enhances the capability of forecasting and automation (Jain & Ramesh, 2020). In (Rehman et al., 2022) Internet of Things has been used to enhance the working capabilities of irrigation systems and fertilization (Rehman et al., 2022). In (Syaza, 2008) an IoT-based system has been proposed to measure the soil moisture to further decide on irrigation. Here a Raspberry Pi-based automated system has been developed to monitor soil moisture. Sensor-based soil analysis is used in (Suhaimi et al., 2021; Swaminathan et al., 2022). Here the NPK value of soil is computed with the help of sensors deployed in the field and afterward, the data gathered is transferred to Google Cloud for future use. The sensor data is analyzed to rectify errors using fuzzy logic interference in (Suhaimi et al., 2021; Swaminathan et al., 2022).

The solution brought out above shows the IoT possibilities that are further utilized. However, the efficiency of the solution can be enhanced with the help of various artificial intelligence techniques. Therefore various deep-learning techniques have been presented here with IoT (Wang et al., 2022). Here in (MINAGRI, 2020), a prototype was invented for smart fertilization to collect real-time information about soil moisture and nutrients. With the help of this system, the farmer gets updates through messages and recommendations about fertilizer to add.

The data-driven Artificial Intelligence and IoT-based solution can be beneficial for the farmers and guide them daily. These solutions can help farmers provide a report of soil quality and guide them to optimize their valuable resources such as water and fertilizer. The data-driven AIoT-based solution for irrigation and fertilization for soil quality will be an innovative solution for farmers and resolve maximum issues related to the wastage of resources. This solution can help the farmer with real-time monitoring of irrigation and fertilization. AIoT systems are empowered with sensors to gather relevant parameters about humidity, temperature and moisture. Afterward, the gathered data was transferred to an AI-powered platform for further analysis to determine the optimal irrigation and fertilization. The collected data is used by an AI-powered platform to assess the quality of the soil. This analysis can be used to identify issues that may affect crop growth and yield. Various applications of AIot in agriculture are listed below:

a. Soil Moisture Monitoring: Artificial Intellience-based IoT system helps to monitor the moisture level of soil and automatically controls the irrigation system based on the outcomes of the monitoring system. Here, sensors are deployed in the field to monitor the moisture of the soil.
b. Precision Irrigation: Irrigation is the most important part of agriculture and AI-based precision irrigation systems help the farmer to control the irrigation based on the data collected through sensor data. The system optimized those factors to control water timing
c. Fertilizer Optimization: Controlled fertilization is really important to produce quality crops Here, the AI-based system collects soil nutrient value based on soil monitoring using sensors. The data collected from sensors are processed with various AI techniques to provide information to farmers.

The benefits of AIoT-based data-driven irrigation and fertilization are listed below:

- Optimization of farming resources can enhance crop yield and quality products.
- AIoT-based systems help the farmer to control irrigation and fertilization.
- This helps farmers take appropriate action against addressed issues.
- Improved sustainability: By using data-driven approaches to optimize irrigation and fertilization practices, farmers can reduce the environmental impact of their operations and improve sustainability.

RELATED WORK

In this paper, the state of the art is further classified into two categories:

a) Irrigation in Agriculture
b) Fertilization in agriculture

The details of the mentioned two categories are laid down in sub-sections:

Irrigation in Agriculture

Several methods have been employed in agriculture for irrigation. The popular four methods are given in Figure 2.

a. Flood irrigation is one of the oldest and most widely used methods of irrigation. The real problem with the method is the huge loss of water due to distribution on the surface of the soil.
b. Spray irrigation is a way of controlling the spray of water on the field. Here, a jet spray is used to spray water like rainfall and distribute it properly. The water distribution is accomplished using a pipe with a pump. The jet spray is used in this technique that's why this is also called a Sprinkle spray.
c. Drip irrigation is a contemporary method also known as trickle irrigation and is famous for the saving of water and soil nutrients. This method allows water to drip slowly over the plant. The slow spray of water must be the reason for more water absorption and reduce the water loss.
d. Tiny water droplets: This method generates fog over fields instead of rain-like spray. This fog stays over the field and comes down as irrigation slowly. Thus, this is also named for irrigation.

Issues Related to Irrigation

This section illustrates the various issues that take place during land irrigation. It has been depicted in Table 1.

Smart Irrigation Techniques

In (Vaishali et al., 2018) a smart automated irrigation system was proposed with the Raspberry Pi and water pump. The system starts irrigation based on the data collected with sensors on fields. The moisture information in the soil acted as a

Figure 2. Types of irrigation

trigger for the Raspberry Pi-based water pump. Based on the value of the moisture in the soil the pump starts the water supply. The pump was directly connected to the smartphone module of the farmer. A smart solution proposed in (Karpagam, 2021) is based on three sensors such as temperature, moisture, and water level. In the system, the microcontroller transmits the control signal to the water pump after processing the signals received from the sensors (Triantafyllou et al., 2019). The threshold value was set and considered to start the irrigation of land.

In (Hwang, 2017) smartphone was used to monitor and control the irrigation process with the help of a wireless sensor-based network. The use of smartphones to control irrigation from remote areas is a challenging task. An intelligent irrigation technique proposed in (Noor & Ramlee, 2021) that controls the water pump based on various soil properties such as moisture, pH value and humidity. The system contains a display unit to show the value of given properties and based on these estimations of these properties threshold values the system regulates the water pump. Although these are better solutions than the manual system but still lack due to the unavailability of artificially intelligent algorithms such as machine learning and deep learning. The AI-based solutions are more data-driven and enable correct prediction of irrigation. Table 2 depicts the state-of-the-art techniques used for fertilization techniques.

Table 1. Irrigation and related problems

Problem	Details
Humid Soil	The water content presents in land soil. (Babaa, 2020; Baravade, 2019; Difallah, 2018a; Imteaj, 2016; Islam, n.d.; Ismail, 2019; Jayadevan, 2020; Joshi, 2019; Kannan et al., 2020; Li, 2020; Rajkumar, 2017; Sachin, 2019; Singh, 2019; Subramani, 2020; Velmurugan, 2020)
Soil temperature	Low and high heated soil (Difallah, 2018a; Difallah, 2018b; Muneeswari, 2017; Rajkumar, 2017; Singh, 2019; Subramani, 2020)
Soil pH	This represents acidic and alkaline value of land Soil (Balaji, 2019; Dasgupta, 2019; Jimenez-Buendıa et al., 2021; Khriji, 2021; Li, 2020; Prakash, 2020; Subramani, 2020)
Soil nu nutrients	Represents soil health (Jimenez-Buendıa et al., 2021; Prakash, 2020; Subramani, 2020)
Air temperature	Details of air heat (Ahmad, n.d.; Akter, 2018; Balaji, 2019; Dasgupta, 2019; Difallah, 2018a; Difallah, 2018b; Hendrawan, 2019; Karpagam, 2020; Laksiri, 2019; Leh, 2019; Manimegalai et al., 2020; Muneeswari, 2017; Naik, 2018; Palconit, 2020; Prakash, 2020; Rajkumar, 2017; Sayed, 2019; Suvra & Sen, n.d.; Zhao, 2017)
Air humidity	Water contents in the air (Ahmad, n.d.; Akter, 2018; Dasgupta, 2019; Difallah, 2018a; Difallah, 2018b; Hendrawan, 2019; Jayadevan, 2020; Joshi, 2019; Karpagam, 2020; Laksiri, 2019; Leh, 2019; Manimegalai et al., 2020; Muneeswari, 2017; Naik, 2018; Palconit, 2020; Pathak et al., 2019; Rajkumar, 2017; Sayed, 2019; Subramani, 2020; Ullah, 2021)
Luminosity	The light intensity within the soil represents Luminosity (Ahmad, n.d.; Difallah, 2018a; Hendrawan, 2019; Imteaj, 2016; Naik, 2018; Pathak et al., 2019; Ullah, 2021).
Precipitation	atmosphere to earth surface waterfall (Ahmad, n.d.; Difallah, 2018a; Jayadevan, 2020; Kannan et al., 2020; Khriji, 2021; Laksiri, 2019; Sayed, 2019; Ullah, 2021)
Leaf wetness	Water on plant surface (Subramani, 2020).
Plant height	Distance from the plant's highest level and the ground level (Dasgupta, 2019).

Fertilization in Agricultural

Fertilization is very important to preserve sufficient quantity and quality of production of any crop. It is necessary to keep the nutrient levels of the food. Mineral fertilization is the best to achieve the needs of the nutrient value and appropriate crop yield. Fertilization is responsible for at least 50% growth in the yield of crops in the past few years. The inappropriate use of mineral fertilization may cause the reason of various issues. The soil fertility can be reduced with the excess use of fertilization and the insufficient amounts of mineral in the soil may cause problems of low yield production. The fertilization impact on soil both ways positive and negative.

Table 2. Fertilization techniques

Reference	Year	Data Pre-Processing	Goal
(Padalalu et al., 2017)	(2017)	–	Estimate the required water
(Perea et al., 2019)	(2019)	–	Irrigation Prediction
(Xie et al., 2017)	(2017)	–	Minimization of irrigation cost
(Kokkonis et al., 2017)	(2017)	Reduce erroneous measurements	Management of irrigation
(Goumopoulos et al., 2014)	(2014)	Data reduction	Management of zone-specific irrigation
(Goldstein et al., 2018)	(2018)	Data cleaning and transformation into the required features	The weekly irrigation plan prediction
(Kang et al., 2000)	(1996a)	–	Management of irrigation
(Hou et al., 2012)	(2009)	–	Management of irrigation (watering time, no watering)
(Perinbam, n.d.)	(2014)	–	Management of irrigation
(Mousa et al., 2014)	(2014)	–	Management of irrigation

In recent times, fertilization was carried out manually and farmers throw fertilizer by hand in the field. This may be the reason for the uneven distribution of fertilization on the land. Therefore, a smart fertilization scheme may help farmers to evenly distribute fertilizers and also improve the productivity of the crops.

Smart Fertilization Techniques

Fertilizers are very important for crops to grow but large amounts of fertilization may harm soil and crop quality. Thus, a measured amount is needed to invest in the field. Initially, fertilization was a very tedious task and was accomplished with manual methods. Manual fertilization leads to various problems related to crop and soil quality. Therefore, researchers and domain scientists have come up with several automated methods to control fertilization and improve crop quality. The invented methods were more focused and ensured the right quantity of fertilization must be applied in the field. A Smart IoT-based system was proposed in (Maheswari et al., 2019) to improve soil fertilization based on the climate conditions. The system collects climate information and based on the information suggests the quantity of fertilization but does not integrate it with the irrigation process to maximize the use of resources. The solution considers the agricultural data stored in the cloud to evaluate the exact situation. Here, big data visual analysis was used to evaluate

Figure 3. Smart fertilization management

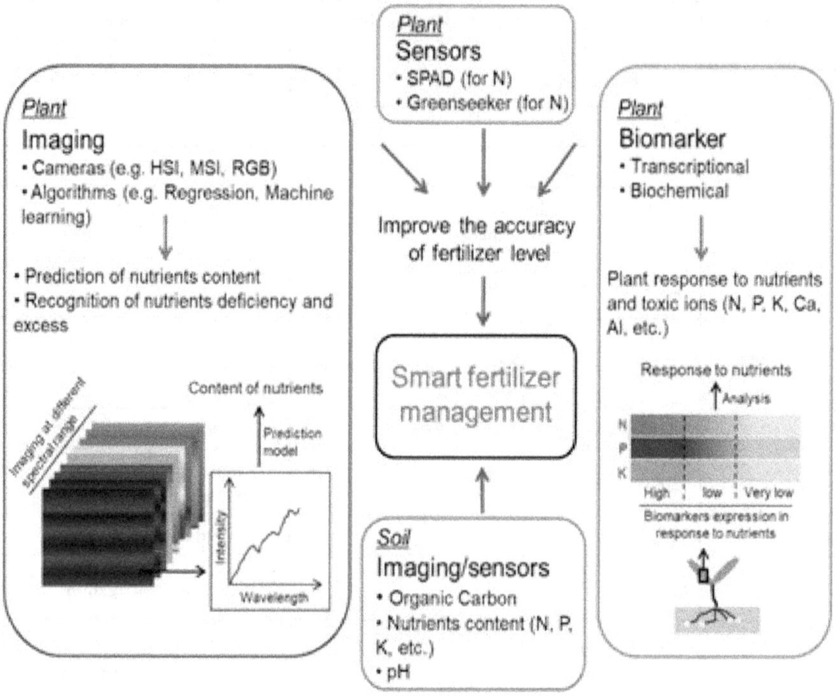

and analyze fertilization analysis for a particular crop and the market availability of fertilizers (Prabha et al., 2018). Here, the mobile application was used to provide information to the farmer, but it was a challenging task as farmers were not equipped with mobile appliances. In (Ullo & Sinha, 2020) a smart system was developed and installed with various sensors to capture information regarding humidity, soil moisture and soil properties. The outcomes of the analysis were stored on the IoT server for further necessary action. But the issue remains the same the solution does not integrate with the fertilization system. A sensor-based technique was proposed in (Mary & Ramakrishnan, 2021) that analyzes the NPK value of soil and based on the outcome provides an appropriate suggestion. The soil color was used to inspect the value of nutrients in the soil, and this was also not associated with the irrigation system. The discussed solution above only resolved a single problem. However, the farmer needs solutions that integrate all the components of smart farming and provide a single solution for the same. Various techniques are listed in Table 3 with the fertilization management system in Figure 3

Several models have been developed and listed in the state of the art to individually tackle irrigation and fertilization-related issues. In this research, the paper describes

Table 3. Smart fertilization details

Imaging Technique	Detection Method/ Algorithm	Nutrients Analyzed Plants		Experimental Setup	Combination of Additional Data	References
	linear-plateau model (Yield- model using VIs)	N-status Maize (canopy)		Field (using airplane)	SPAD data	(Maresma et al., 2018) 2018
	SAM (prediction model)	N-content Orange(leaf)		Field (using UAV)	Spectroradiometer data (Ground sensor)	(Osco et al., 2019) 2019b
	Stepwise-MLR (prediction model)	N-content Rice (canopy)		Field (using UAV)	Hyperspectral data (Ground sensor)	(Zheng et al., 2018) 2018
	LRM (prediction model)	N-use efficiency for Winter Wheat (canopy) variety		Field (Using UAV)	No	(Herrero et al., 2020) 2020
MSI	LRM (prediction model)	N-content	Sugarcane(leaf)	Field (Using UAV)	No	(Shendryk et al., 2020) et al. 2020
	MLR (prediction model with VIs)	N-status	Oilseed rape (canopy)	Field (using UAV)	Hyperspectral data (Ground Sensor)	(Liu et al., 2018). 2018
	LRM (evaluate the performance of the VIs)	N-status	Winter Wheat (canopy)	Field (using UAV)	No	(Lu, Wang, Zhang, Li, Yao, Tian, Zhu, Cao, Baret, Liu, & Cheng, 2019) 2019
	LRM (estimation model)	N-content	Winter Wheat (canopy)	Filed (using UAV)	No	(Geipel et al., 2016) 2016
	ELR (estimation model)	N-management	Soybean (canopy)	Filed (using UAV)	RGB image &Thermal image	(Maimaitijiang et al., 2017) 2017
	MLR (prediction model)	N-deficiency	Cucumber (leaf)	Greenhouse & Lab	No	(Ji-Yong, Xiao-Bo, Jie-Wen, Kai-Liang, Zheng-Wei, Xiao-Wei, De-Tao, & Holmes, 2012) 2012
	RF (wavelength selection), PLSR (prediction model)	N-content	Pepper (leaf-stem-root)	Greenhouse & Lab	No	(Zhang et al., 2014). 2014
	SG-PLS (prediction model for N), N & P-contents PLS(prediction model for P)		Citrus (leaf)	Greenhouse & Lab	No	(Liu et al., 2015) 2015
	LNC-indices (prediction model)	N-content	Rice (canopy)	Field & Lab	No	(Moharana & Dutta, 2016) 2016
	PLSR (prediction model)	N-content (high accuracy)P,K, S-contents (less accuracy)	Maize (leaf), Soybean (leaf)	Lab	No	(Pandey et al., 2017). 2017
HSI	PLSR (prediction model)	N-content	Spinach (canopy)	Greenhouse & Lab	No	(Corti et al., 2017) 2017
	SAE-FNN (Predicted model)	N-content	Oilseed rape-leaf	Field & Lab	No	(Yu et al., 2018) 2018
	PLSR (prediction model)	N-status	Wheat (leaf)	Lab	No	(Bruning et al., 2019). 2019
	Normalized difference chlorophyll index	N-use efficiency for Wheat (leaf) variety		Lab	RGB	(Banerjee et al., 2020) 2020
	MLR (prediction model)	N-content	Apple (leaf & canopy)	Lab and field	No	(Ye et al., 2020) 2020
	PCA(Image feature evaluation)	N-status	Barley (leaf)	Greenhouse & Lab	SPAD data	(Pagola et al., 2009) 2009

Table 3. Continued

Imaging Technique	Detection Method/ Algorithm	Nutrients Analyzed Plants		Experimental Setup	Combination of Additional Data	References
	LRM (RGB component analysis)	N-status	Wheat (canopy)	Field	GreenSeeker data	(Li et al., 2010) 2010
	LRM (RGB component analysis)	N-status	Tomato (leaf)	Greenhouse & Lab	SAPD data	(Mercado-Luna et al., 2010) 2010
	Correlation modeling	N-deficiency	Rubber Tree (leaf)	Lab	No	(Lu, Wang, Zhang, Li, Yao, Tian, Zhu, Cao, Baret, Liu, & Cheng, 2019) 2012

a model that tackles both irrigation and fertilization issues in a go. The proposed model resolves issues related to fertilization and irrigation.

AIOT-BASED SMART FARMING SYSTEM

AIoT-based smart farming system consists of four stages and each stage is defined with a specific task. The smart farming system consists of four stages based on OSI architecture as shown in Figure 4.

A. Data collection
B. Data Access
C. Data Storage and Interpretation
D. Data Presentation and Visualization

Data Gathering

In this stage, the data is collected from the field with the help of Sensors deployed in the agriculture field. The agro sensors can be deployed on the crop or within the soil as shown in the figure. Different sensors are used in the field to collect information about the soil parameters such as moisture content and nutrient values. Here, the sensors are deployed in the field at various locations to generate information. The deployed sensors formed a wireless network to generate information. The sensor generates a signal after sensing data and converts it into output signals. The deployed sensors collect macroscopic scales (temperature, electrical power, pressure, NPK, humidity) and convert them into electronic signals with higher accuracy. The following sensors are most commands that are used in the field to collect the data.

Figure 4. AIoT-based smart farming system architecture

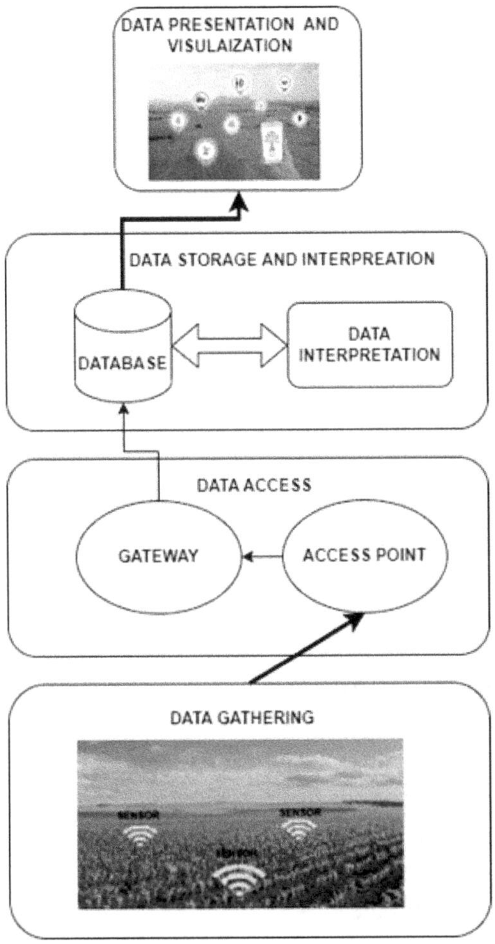

1. Optical Sensors: Sensors are used to sense light reflection from soil or plants. The optical sensors can collect information about the color, temperature and moisture content of soil.
2. Weather measuring sensors: This sensor is used to collect information about weather such as the air pressure, humidity level, air direction and flow.
3. NPK Sensors: The soil nutrient values such as nitrogen, potassium and phosphorus can be estimated with the help of NPK sensors. This is used to calculate the level of NPK in soil and these values can be used to estimate the soil fertility. The sensors used for this task can be used for any type of soil because the sensors are protected with high-quality material. NPK sensors are buried into the soil to generate valuable information.

Several types of sensors are used to get estimates about humidity levels, nutrient levels and other information for smart farming. The collected data is stored in the database through a network established through an access point and gateway.

Data Access

In studies, it has been analyzed that no proper communicable ways or protocols are available in agriculture. The prime objective here is to establish a connection with data storage and store data generated through the previous layer. The main task is to establish proper communication through efficient services. Here, once the data is received from the previous stage it should be shared with a gateway for further course of action. The following devices are further used here to establish proper communication as listed below:

a) Wi-Fi is a wireless protocol based on IEEE standards and enables communication across WLAN at different frequencies.
b) GPRS and ETSI are used for communication and use the latest mobile technologies for the fastest transmission.

Data Storage and Interpretation

The most important stage of smart farming architecture is where the collected data has been stored and interpreted for further analysis. The complete working stage is carried out in two phases:

a) Data cleansing and Storage
b) Data Interpretation

Data Cleansing and Storage

Here, the sensor-generated data has been filtered and cleansed for any discrepancy before use. Several programming constructs, open-source tools and data engineering tools are available to accomplish this task. Here in this research, the task is carried out with a programming construct. The filtered data after removing anomalies has been stored in the database for further processing.

Data Interpretation

In this phase, the raw data from the database has been used to generate various hidden patterns using several artificial intelligence methods and techniques. Here, various machine and deep learning techniques have been employed to generate hidden patterns and based on these patterns the smart farming system provides information to the farmers. The generated patterns are stored in a database for further analysis. Smart farming systems use various image processing to process image-based and interpret the same with the help of artificial intelligence-based techniques. In the next stage, the smart farming system provides a visual display of information based on the pattern generated.

Presentation and Visualization

Smart farming systems provide ease to users with the help of mobile apps. The farming system provides visual data and information regarding fertilization and irrigation on users' or farmers' smart mobiles with the help of the mobile app. Based on the real-time information the farmer will get the real-time information about irrigation and fertilization situation.

Proposed Algorithm

Here in this work, the machine and deep learning algorithms have been used to predict irrigation and fertilization prediction based on available data. Various deep learning models such as Artificial Neural Network(ANN), Long Short Term Memory(LSTM), Recurrent neural network (RNN) and Bi-directional Long Short Term Memory (Bi-LSTM) has been employed to show the prediction of irrigation and fertilization. In AIoT proposed system, a hybrid model have been employed to predict the value of fertilization and irrigation. The hybridization has been achieved with ANN and Bi-LSTM. In analysis, it has been observed that the hybrid model shows better prediction accuracy than other deep learning. The stepwise process has been laid down below in steps and Figure 5:

1. Data generated from various sensors has been captured and checked for redundancy. The missing value from the data has been rectified and replaced with the help of various operations. The accurate data that contains humidity value, temperature and NPK values will be considered for further processing.
2. The complete data has been divided into training-testing ratios such as 60:40, 70:40 and 80:20. The results of the experiment have been laid down in the result and analysis section. The available portion of data has been used to train the model.
3. The hybrid model has been used in the system to predict fertilization and irrigation values. The structure of the hybrid model is laid down below:

Figure 5. Flow of methodology

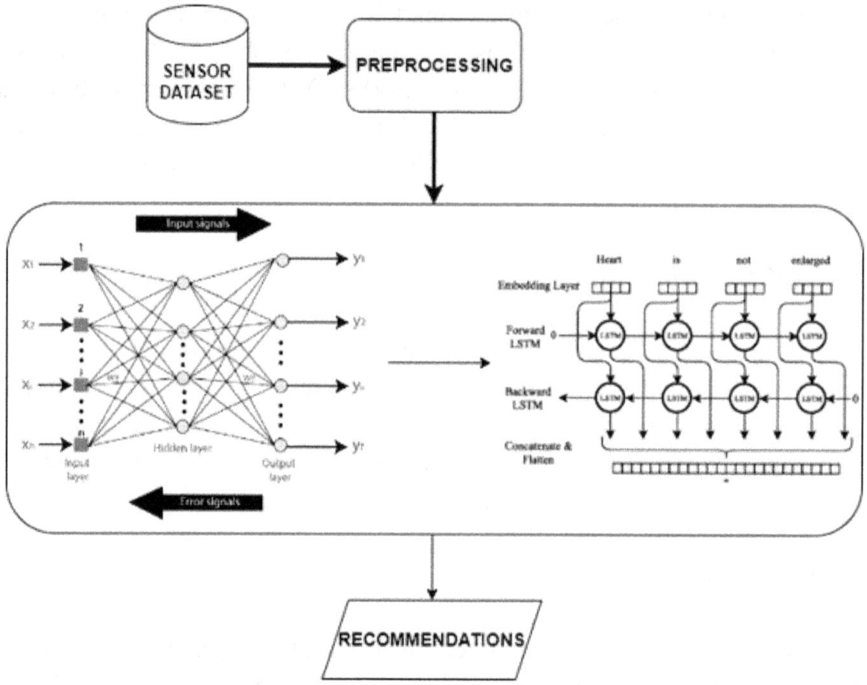

The model has been elaborated with following equations:

$$I_i = w_{ij} * x_{ij} + u_i$$

Where:

I_i : Input weighted with i weight

w_{ij} : weight for neuron i o j

x_{ij} : output fron i to j neuron

u_i : *baseline input*

Figure 6. Accuracy comparison of different DL algorithms

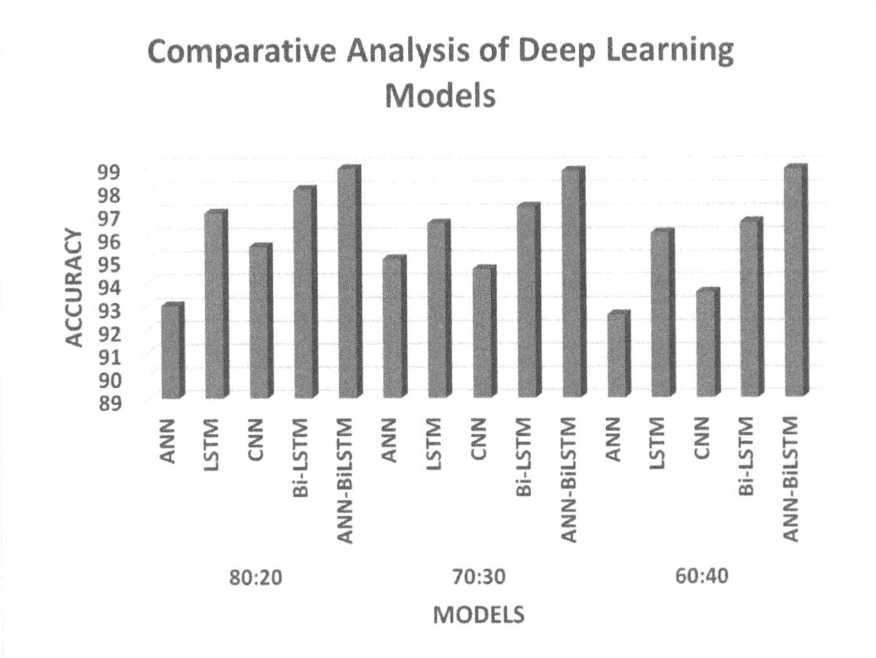

$$Y_i = \sum_{j=1}^{n} {}_{i=1} w_{ij} * x_{ij}$$

Bi-LSTM representation is as follows:

$$v_i = \mathrm{B}\left(Y_{1:n} : \mathrm{i}\right)$$

RESULT AND DISCUSSION

This section covers the performance analysis of various deep learning models such as CNN, LSTM, Bi-LSTM and ANN-BiLSTM models. Sensors generate huge amounts of data, that's why deep learning techniques were used for prediction. The complete implementation has been accomplished with Python programming and tested on the soil nutrient dataset. The model was trained with fertilization, cultivation, and nutrient datasets. The complete experiment was performed with different sampling data such as 80:20, 70:30 and 60:40. The accuracy rate of all the deep learning models is depicted in Table 4 and Figure 6.

Finally, the observation is that the hybrid model outperformed and showed a prediction accuracy of 98.75%.

CONCLUSION

This research proposed a new AIoT-based smart system based on recent technologies. This system provides the latest information on irrigation and fertilization to farmers. Smart system will collect irrigation and fertilization data through various sensors deployed in the farms and the collected data will be used to assist farmers in further planning. The smart system was embedded with deep learning technology to predict fertilization and irrigation requirements based on real-time data. In the future, the system will deployed with disease and weed detection.

REFERENCES

Abdelmoamen Ahmed, A., Al Omari, S., Awal, R., Fares, A., & Chouikha, M. (2021). A distributed system for supporting smart irrigation using internet of things technology. *Engineering Reports*, *3*(7), e12352. doi:10.1002/eng2.12352

Ahmad. (n.d.). *Client server smart irrigation system*. Academic Press.

Ahmed, I., & Sikder, M. (2020). IoT based smart irrigation system at university of Chittagong, Bangladesh. *International Journal of Computer Applications*, *176*(15), 39–45. doi:10.5120/ijca2020920097

Aker, J. C., Ghosh, I., & Burrell, J. (2016, November). The promise (and pitfalls) of ICT for agriculture initiatives. *Agricultural Economics*, *47*(S1, no. S1), 35–48. doi:10.1111/agec.12301

Akter, S. (2018). Developing a smart irrigation system using arduino. *International Journal of Research Studies in Science, Engineering and Technology*, *6*(1), 31–39.

Al Nahian. (2021). *IoT based soil monitoring and automatic irrigation system*. Academic Press.

Ayaz, M., Ammad-Uddin, M., Sharif, Z., Mansour, A., & Aggoune, E.-H. M. (2019). Internet-of-things (iot)-based smart agriculture: Toward making the fields talk. *IEEE Access : Practical Innovations, Open Solutions*, *7*, 129551–129583. doi:10.1109/ACCESS.2019.2932609

Babaa, S. E. (2020). Smart irrigation system using arduino with solar power. *International Journal of Engineering Research & Technology (Ahmedabad)*, *9*(05), 7.

Balaji, V. (2019). Smart irrigation system using iot and image processing. *International Journal of Engineering and Advanced Technology*, *8*(6s, no. 6S), 115–120. doi:10.35940/ijeat.F1024.0886S19

Banerjee, B., Joshi, S., Thoday-Kennedy, E., Pasam, R. K., Tibbits, J., Hayden, M., Spangenberg, G., & Kant, S. (2020). High-throughput Phenotyping Using Digital and Hyperspectral Imaging-derived Biomarkers for Genotypic Nitrogen Response. *Journal of Experimental Botany*, *71*(15), 4604–4615. doi:10.1093/jxb/eraa143 PMID:32185382

Baravade, D. (2019). Study paper on smart irrigation system. *International Journal of Engineering Sciences & Research Technology*, *8*(4), 99–102.

Bruning, B., Liu, H., Brien, C., Berger, B., Lewis, M., & Garnett, T. (2019). The Development of Hyperspectral Distribution Maps to Predict the Content and Distribution of Nitrogen and Water in Wheat (Triticum Aestivum). *Frontiers in Plant Science*, *10*, 1380. Advance online publication. doi:10.3389/fpls.2019.01380 PMID:31737009

Corti, M., Gallina, P. M., Cavalli, D., & Cabassi, G. (2017). Hyperspectral Imaging of Spinach Canopy under Combined Water and Nitrogen Stress to Estimate Biomass, Water, and Nitrogen Content. *Biosystems Engineering*, *158*, 38–50. doi:10.1016/j.biosystemseng.2017.03.006

Dasgupta, A. (2019). Smart irrigation: Iot-based irrigation monitoring system. *Proceedings of international ethical hacking conference 2018*, 395–403. 10.1007/978-981-13-1544-2_32

Difallah, W. (2018a). Design of a solar powered smart irrigation system (spsis) using wsn as an iot device. *Proceedings of the 2018 International Conference on Software Engineering and Information Management*, 124–128. 10.1145/3178461.3178482

Difallah, W. (2018b). Intelligent irrigation management system. Energy, 9(9). doi:10.14569/IJACSA.2018.090954

Geipel, J., Link, J., Wirwahn, J. A., & Claupein, W. (2016). A Programmable Aerial Multispectral Camera System for In-season Crop Biomass and Nitrogen Content Estimation. *Agriculture*, *6*(1), 4. doi:10.3390/agriculture6010004

Goldstein, A., Fink, L., Meitin, A., Bohadana, S., Lutenberg, O., & Ravid, G. (2018). Applying machine learning on sensor data for irrigation recommendations: Revealing the agronomist's tacit knowledge. *Precision Agriculture, 19*(3), 421–444. doi:10.1007/s11119-017-9527-4

Goumopoulos, C., O'Flynn, B., & Kameas, A. (2014). Automated zone-specific irrigation with wireless sensor/actuator network and adaptable decision support. *Computers and Electronics in Agriculture, 105*, 20–33. doi:10.1016/j.compag.2014.03.012

Hassan, A. (2021). A wirelessly controlled robot-based smart irrigation system by exploiting arduino. *Journal of Robotics and Control, 2*(1), 29–34.

Hendrawan, I. N. R. (2019). Fuzzy based internet of things irrigation system. *1st International Conference on Cybernetics and Intelligent System (ICORIS),* 1, 146–150. 10.1109/ICORIS.2019.8874900

Herrero, H., Waylen, P., Southworth, J., Khatami, R., Yang, D., & Child, B. (2020). A Healthy Park Needs Healthy Vegetation: The Story of Gorongosa National Park in the 21st Century. *Remote Sensing (Basel), 12*(3), 476. doi:10.3390/rs12030476

Hou, H., Peng, S., Xu, J., Yang, S., & Mao, Z. (2012). *Seasonal variations of CH4 and N2O emissions in response to water management of paddy fields located in Southeast China*. Academic Press.

Hwang, S. (2017). Monitoring and Controlling System for an IoT Based Smart Home. *Int. J. Control Autom., 10*(2), 339–348. doi:10.14257/ijca.2017.10.2.28

Imteaj, A. (2016). Iot based autonomous percipient irrigation system using raspberry pi. *19th International Conference on Computer and Information Technology (ICCIT),* 563–568. 10.1109/ICCITECHN.2016.7860260

Islam. (n.d.). *Iot based smart irrigation monitoring & controlling system in agriculture*. Academic Press.

Ismail, N. (2019). *Smart irrigation system based on internet of things (iot). Journal of Physics: Conference Series*, 1339.

Jain, S., & Ramesh, D. (2020). Machine Learning convergence for weather based crop selection. *2020 IEEE Int. Students' Conf. Electr. Electron. Comput. Sci. SCEECS.*

Jayadevan, J. (2020). A novel architecture for internet of things in precision agriculture. *International Journal of Applied Engineering Research, 15*(3), 204–211.

Ji-Yong, S., Xiao-Bo, Z., Jie-Wen, Z., Kai-Liang, W., Zheng-Wei, C., Xiao-Wei, H., De-Tao, Z., & Holmes, M. (2012). Nondestructive Diagnostics of Nitrogen Deficiency by Cucumber Leaf Chlorophyll Distribution Map Based on near Infrared Hyperspectral Imaging. *Scientia Horticulturae, 138*, 190–197. doi:10.1016/j.scienta.2012.02.024

Jimenez-Buendıa, M., Soto-Valles, F., Blaya-Ros, P. J., Toledo-Moreo, A., Domingo-Miguel, R., & Torres-Sánchez, R. (2021). High-density wi-fi based sensor network for efficient irrigation management in precision agriculture. *Applied Sciences (Basel, Switzerland), 11*(4), 1628. doi:10.3390/app11041628

Joshi, D. (2019). Smart irrigation system using iot for farmers. *International Journal of Emerging Technologies and Innovative Research, 6*(1), 207–209.

Kang, S., Liang, Z., Pan, Y., Shi, P., & Zhang, J. (2000). Alternate furrow irrigation for maize production in an arid area. *Agricultural Water Management, 45*(3), 267–274. doi:10.1016/S0378-3774(00)00072-X

Kannan, K., Kumar, N. S., Logith, E., Kumar, R. M., & Prakash, O. S. (2020). An internet of things based smart irrigation using solenoid valve. *International Journal of Recent Technology and Engineering, 9*(1), 2018–2023. doi:10.35940/ijrte.A1979.059120

Kapitanova, K., & Son, S. H. (2012). Machine learning basics. Intell. Sens. Networks Integr. Sens. Networks, Signal Process. Mach. Learn., Ml, 3–29. doi:10.1201/b14300-2

Karar, M. E. (2021). *A pilot study of smart agricultural irrigation using unmanned aerial vehicles and iot-based cloud system*. arXiv preprint arXiv:2101.01851.

Karpagam, J. (2021). *7th International Conference on Advanced Computing and Communication Systems, ICACCS 2021*. Academic Press.

Karpagam, J. (2020). Smart irrigation system using iot. *6th International Conference on Advanced Computing and Communication Systems (ICACCS)*, 1292–1295.

Kathuria, S. (2018). E-krishi: An iot based smart irrigation system. *International Journal of Electrical, Electronics and Data Communication, 6*(6), 8–13.

Khriji, S. (2021). *Precision irrigation: an iot-enabled wireless sensor network for smart irrigation systems. In Women in precision agriculture*. Springer.

Kokkonis, G., Kontogiannis, S., & Tomtsis, D. (2017). A smart IoT fuzzy irrigation system. *Power, 100*(63), 25.

Laksiri, H. (2019). Design and optimization of iot based smart irrigation system in sri lanka. *14th Conference on Industrial and Information Systems (ICIIS),* 198–202. 10.1109/ICIIS47346.2019.9063272

Leh, N. A. M. (2019). Smart irrigation system using internet of things. *9th International Conference on System Engineering and Technology (ICSET),* 96–101. 10.1109/ICSEngT.2019.8906497

Li, W. (2020). *Review of sensor network-based irrigation systems using iot and remote sensing*. Advances in Meteorology.

Li, Y., Chen, D., Walker, C. N., & Angus, J. F. (2010). Estimating the Nitrogen Status of Crops Using a Digital Camera. *Field Crops Research, 118*(3), 221–227. doi:10.1016/j.fcr.2010.05.011

Liu, S., Li, L., Gao, W., Zhang, Y., Liu, Y., Wang, S., & Lu, J. (2018). Diagnosis of Nitrogen Status in Winter Oilseed Rape (Brassica Napus L.) Using In-situ Hyperspectral Data and Unmanned Aerial Vehicle (UAV) Multispectral Images. *Computers and Electronics in Agriculture*, *151*, 185–195. doi:10.1016/j.compag.2018.05.026

Liu, Y., Lyu, Q., He, S., Yi, S., Liu, X., Xie, R., Zheng, Y., & Deng, L. (2015). Prediction of Nitrogen and Phosphorus Contents in Citrus Leaves Based on Hyperspectral Imaging. *International Journal of Agricultural and Biological Engineering*, *8*, 80–88. doi:10.3965/j.ijabe.20150802.1464

Lu, N., Wang, W., Zhang, Q., Li, D., Yao, X., Tian, Y., Zhu, Y., Cao, W., Baret, F., Liu, S., & Cheng, T. (2019). Estimation of Nitrogen Nutrition Status in Winter Wheat from Unmanned Aerial Vehicle Based Multi-angular Multispectral Imagery. *Frontiers in Plant Science*, *10*, 1601. doi:10.3389/fpls.2019.01601 PMID:31921250

Maheswari, R., Azath, H., Sharmila, P., & Sheeba Rani Gnanamalar, S. (2019). Smart Village: Solar Based Smart Agriculture with IoT Enabled for Climatic Change and Fertilization of Soil. *2019 IEEE 5th Int. Conf. Mechatronics Syst. Robot. ICMSR 2019*, 102–105. 10.1109/ICMSR.2019.8835454

Maimaitijiang, M., Ghulam, A., Sidike, P., Hartling, S., Maimaitiyiming, M., Peterson, K., Shavers, E., Fishman, J., Peterson, J., Kadam, S., Burken, J., & Fritschi, F. (2017). Unmanned Aerial System (Uas)-based Phenotyping of Soybean Using Multi-sensor Data Fusion and Extreme Learning Machine. *ISPRS Journal of Photogrammetry and Remote Sensing*, *134*, 43–58. doi:10.1016/j.isprsjprs.2017.10.011

Manimegalai, V., Judy, A. L., Gayathri, A., Ashadevi, S., & Mohanapriya, V. (2020). Smart irrigation system with monitoring and controlling using iot. *International Journal of Engineering and Advanced Technology*, *9*(4), 1373–1376. doi:10.35940/ijeat.C6586.049420

Maresma, Á., Lloveras, J., & Martínez-Casasnovas, J. A. (2018). Use of multispectral airborne images to improve in-season nitrogen management, predict grain yield and estimate economic return of maize in irrigated high yielding environments. *Remote Sensing (Basel)*, *10*(4), 543. doi:10.3390/rs10040543

Mary, D. L., & Ramakrishnan, M. (2021). A Novel Approach to Optimize Water and Fertilizers in Agriculture using IoT. *Int. J. Cybern. Informatics*, *10*(2), 57–64. doi:10.5121/ijci.2021.100207

Mercado-Luna, A., Rico-García, E., Lara-Herrera, A., Soto-Zarazúa, G., Ocampo-Velázquez, R., Guevara-González, R., Herrera-Ruiz, G., & Torres-Pacheco, I. (2010). Nitrogen Determination on Tomato (Lycopersicon Esculentum Mill.) Seedlings by Color Image Analysis (RGB). *African Journal of Biotechnology*, *9*, 5326–5332.

MINAGRI. (2020). *Minagri Annu. Rep. 2019-2020.* Author.

Mishra, D., Abbas, A., Pande, T., Pandey, A. K., Agrawal, K. K., & Yadav, R. S. (2019). Smart agriculture system using IoT. *ACM Int. Conf. Proceeding Ser.*

Moharana & Dutta. (2016). Spatial Variability of Chlorophyll and Nitrogen Content of Rice from Hyperspectral Imagery. *ISPRS J. Photogramm. Remote Sens, 122*, 17–29. doi:.isprsjprs.2016.09.002 doi:10.1016/j

Mousa, A. K., Croock, M. S., & Abdullah, M. N. (2014). Fuzzy based decision support model for irrigation system management. *International Journal of Computer Applications*, *104*(9). Advance online publication. doi:10.5120/18230-9177

Muneeswari. (2017). Smart irrigation system using iot approach. *International Journal of Engineering Research & Science.*

Munir, M. S. (2021). *Intelligent and smart irrigation system using edge computing and iot.* Complexity.

Naik, P. (2018). Automation of irrigation system using iot. International Journal of Engineering and Manufacturing Science, 8(1), 77–88.

Noor, M. Z. M., & Ramlee, R. A. (2021). Performances Analysis of IoT Based Smart Greenhouse System. *Int. J. Electr. Eng. Appl. Sci.*, *4*(2), 1–8.

Orn, D., Duan, L., Liang, Y., Siy, H., & Subramaniam, M. (2020). Agro-AI Education: Artificial Intelligence for Future Farmers. *SIGITE 2020 - Proc. 21st Annu. Conf. Inf. Technol. Educ.*, 54–57. 10.1145/3368308.3415457

Osco, L. P., Ramos, A. P. M., Pereira, D. R., Éas, M., Imai, N. N., Matsubara, E. T., & Estrabis, N. (2019). Predicting Canopy Nitrogen Content in Citrus-trees Using Random Forest Algorithm Associated to Spectral Vegetation Indices from UAV-Imagery. *Remote Sensing (Basel)*, *11*(24), 2925. doi:10.3390/rs11242925

Padalalu, P., Mahajan, S., Dabir, K., Mitkar, S., & Javale, D. (2017). Smart water dripping system for agriculture/farming. *2017 2nd International Conference for Convergence in Technology (I2CT),* 659–662. 10.1109/I2CT.2017.8226212

Pagola, Ortiz, Irigoyen, Bustince, Barrenechea, Aparicio-Tejo, Lamsfus, & Lasa. (2009). New Method to Assess Barley Nitrogen Nutrition Status Based on Image Colour Analysis: Comparison with SPAD-502. *Comput. Electron. Agric, 65*(2), 213–218. doi:.compag.2008.10.003 doi:10.1016/j

Palconit, M. G. B. (2020). Iot-based precision irrigation system for eggplant and tomato. *9th Int. Symp. Comput. Intell. Ind. Appl.*

Pandey, P., Ge, Y., Stoerger, V., & Schnable, J. C. (2017). High Throughput in Vivo Analysis of Plant Leaf Chemical Properties Using Hyperspectral Imaging. *Frontiers in Plant Science, 8*, 1348. doi:10.3389/fpls.2017.01348 PMID:28824683

Pathak, A., AmazUddin, M., Abedin, M. J., Andersson, K., Mustafa, R., & Hossain, M. S. (2019). Iot based smart system to support agricultural parameters: A case study. *Procedia Computer Science*, *155*, 648–653. doi:10.1016/j.procs.2019.08.092

Perea, R. G., Poyato, E. C., Montesinos, P., & Díaz, J. A. R. (2019). Prediction of irrigation event occurrence at farm level using optimal decision trees. *Computers and Electronics in Agriculture.*

Perinbam. (n.d.). Automatic Irrigation System using Fuzzy Logic. *Hemosphere, 89*(7), 884-892.

Podder. (2021). Iot based smart agrotech system for verification of urban farming parameters. *Microprocessors and Microsystems, 82*.

Prabha, R., Sinitambirivoutin, E., Passelaigue, F., & Ramesh, M. V. (2018). Design and Development of an IoT Based Smart Irrigation and Fertilization System for Chilli Farming. *2018 Int. Conf. Wirel. Commun. Signal Process. Networking, WiSPNET 2018*, 1–7. 10.1109/WiSPNET.2018.8538568

Pradeep, K., & Balasundaram, A. (2021). Iot based smart irrigation for agricultural fields. *Annals of the Romanian Society for Cell Biology*, 2000–2009.

Prakash, B. R. (2020). Super smart irrigation system using internet of things. *2020 7th International Conference on Smart Structures and Systems (ICSSS),* 1–5. 10.1109/ICSSS49621.2020.9202361

Rajkumar, M. N. (2017). Intelligent irrigation systema^C"an iot based approach. *International Conference on Innovations in Green Energy and Healthcare Technologies (IGEHT),* 1–5.

Rawal, S. (2017). Iot based smart irrigation system. *International Journal of Computer Applications, 159*(8), 7–11. doi:10.5120/ijca2017913001

Rehman, A., Saba, T., Kashif, M., Fati, S. M., Bahaj, S. A., & Chaudhry, H. (2022). A Revisit of Internet of Things Technologies for Monitoring and Control Strategies in Smart Agriculture. *Agronomy (Basel), 12*(1), 1–21. doi:10.3390/agronomy12010127

Sachin, P. (2019). Iot based smart irrigation system. *International Journal of Advanced Research in Electrical, Electronics and Instrumentation Engineering, 8*(5), 1604–1608.

Sayed, M. A. (2019). An iot based robotic system for irrigation notifier. *International Conference on Robotics, Automation, Artificial-intelligence and Internet-of-Things (RAAICON),* 77–80. 10.1109/RAAICON48939.2019.38

Shendryk, Y., Sofonia, J., Garrard, R., Rist, Y., Skocaj, D., & Thorburn, P. (2020). Fine-scale Prediction of Biomass and Leaf Nitrogen Content in Sugarcane Using UAV LiDAR and Multispectral Imaging. *International Journal of Applied Earth Observation and Geoinformation, 92*, 102177. doi:10.1016/j.jag.2020.102177

Shufian, A. (2019). Smart irrigation system with solar power and gsm technology. *5th International Conference on Advances in Electrical Engineering (ICAEE),* 301–305. 10.1109/ICAEE48663.2019.8975634

Singh, K. (2019). Iot based approach for smart irrigation system suited to multiple crop cultivation. *International Journal of Engineering Research & Technology (Ahmedabad), 12*, 357–363.

Singh, K. (2021). *Design of a low-cost sensor-based iot system for smart irrigation. In Applications in Ubiquitous Computing*. Springer.

Subramani, C. (2020). *Iot-based smart irrigation system. In Cognitive Informatics and Soft Computing*. Springer.

Sudharshan, N., Karthik, A. V. S. K., Kiran, J. S. S., & Geetha, S. (2019). Renewable energy based smart irrigation system. *Procedia Computer Science, 165*, 615–623. doi:10.1016/j.procs.2020.01.055

Suhaimi, A. F., Yaakob, N., Saad, S. A., & Azami, K. (2021). *IoT Based Smart Agriculture Monitoring*. Automation and Intrusion Detection System IoT Based Smart Agriculture Monitoring, Automation and Intrusion Detection System.

Supriya, B. (2021). Iot based remote smart irrigation system. *International Journal of Modern Agriculture, 10*(2), 3805–3811.

Suvra & Sen. (n.d.). *Iot based automated irrigation and smart agriculture monitoring system*. Academic Press.

Swaminathan, Palani, Kotecha, Kumar, & V. (2022). IoT Driven Artificial Intelligence Technique for Fertilizer Recommendation Model. *IEEE Consum. Electron. Mag*.

Syaza, N. B. I. (2008). Smart Home Garden Irrigation System With Raspberry Pi. *IEEE, 16*(June), 24.

Thakur, D., Kumar, Y., & Vijendra, S. (2020). Smart irrigation and intrusions detection in agricultural fields using iot. *Procedia Computer Science, 167*, 154–162. doi:10.1016/j.procs.2020.03.193

Triantafyllou, Sarigiannidis, & Bibi. (2019). Precision agriculture: A remote sensing monitoring system architecture. *Inf., 10*(11).

Ullah, R. (2021). *Eewmp: an iot-based energy-efficient water management platform for smart irrigation*. Scientific Programming.

Ullo, S. L., & Sinha, G. R. (2020). Advances in smart environment monitoring systems using iot and sensors. *Sensors (Basel), 20*(11), 1–18. doi:10.3390/s20113113 PMID:32486411

Vaishali, S., Suraj, S., Vignesh, G., Dhivya, S., & Udhayakumar, S. (2018). Mobile integrated smart irrigation management and monitoring system using IOT. *Proc. 2017 IEEE Int. Conf. Commun. Signal Process. ICCSP 2017,* 2164–2167.

Velmurugan. (2020). *An iot based smart irrigation system using soil moisture and weather prediction*. Academic Press.

Wang, D., Cao, W., Zhang, F., Li, Z., Xu, S., & Wu, X. (2022). A Review of Deep Learning in Multiscale Agricultural Sensing. *Remote Sensing (Basel), 14*(3), 559. doi:10.3390/rs14030559

Xie, T., Huang, Z., Chi, Z., & Zhu, T. (2017). Minimizing amortized cost of the on-demand irrigation system in smart farms. *3rd International Workshop on Cyber-Physical Systems for Smart Water Networks*, 43–46. 10.1145/3055366.3055370

Ye, X., Abe, S., & Zhang, S. (2020). Estimation and Mapping of Nitrogen Content in Apple Trees at Leaf and Canopy Levels Using Hyperspectral Imaging. *Precision Agriculture*, *21*(1), 198–225. doi:10.1007/s11119-019-09661-x

Yu, X., Lu, H., & Liu, Q. (2018). Deep-learning-based Regression Model and Hyperspectral Imaging for Rapid Detection of Nitrogen Concentration in Oilseed Rape (Brassica Napus L.) Leaf. *Chemometrics and Intelligent Laboratory Systems*, *172*, 188–193. doi:10.1016/j.chemolab.2017.12.010

Zhang, D., Song, H., Cheng, H., Hao, D., Wang, H., Kan, G., Jin, H., & Yu, D. (2014). The Acid Phosphatase-encoding Gene GmACP1 Contributes to Soybean Tolerance to Low-phosphorus Stress. *PLOS Genetics*, *10*(1), 1. doi:10.1371/journal.pgen.1004061 PMID:24391523

Zhao, W. (2017). *Design and implementation of smart irrigation system based on lora. In IEEE Globecom Workshops (GC Wkshps)*. IEEE.

Zheng, H., Cheng, T., Li, D., Yao, X., Tian, Y., Cao, W., & Zhu, Y. (2018). Combining Unmanned Aerial Vehicle (Uav)-based Multispectral Imagery and Ground-based Hyperspectral Data for Plant Nitrogen Concentration Estimation in Rice. *Frontiers in Plant Science*, *9*, 936. doi:10.3389/fpls.2018.00936 PMID:30034405

Chapter 3
AIoT Integration Advancements and Challenges in Smart Sensing Technologies for Smart Devices

V. Dankan Gowda
https://orcid.org/0000-0003-0724-0333
B.M.S. Institute of Technology and Management, Bangalore, India

Mandeep Kaur
https://orcid.org/0000-0003-0774-1808
Punjabi University, India

D. Srinivas
https://orcid.org/0000-0002-6446-6649
SR University, India

K. D. V. Prasad
https://orcid.org/0000-0001-9921-476X
Symbiosis Institute of Business Management, Symbiosis International University, India

R. Shekhar
Alliance University, Bangalore, India

ABSTRACT

Artificial intelligence in things (AIoT) has revolutionized the capabilities and features of smart gadgets. Technology advancements in smart sensing have allowed for the seamless integration of AI and IoT, increasing the general intelligence and efficacy of smart devices. This chapter looks at smart sensing technology's

DOI: 10.4018/979-8-3693-0786-1.ch003

progress and challenges in the context of AIoT integration. The study begins with a brief introduction of AIoT and its significance in the smart device industry. It then delves into the numerous smart sensing technologies that aid in bringing AI and IoT together, such as environmental sensors, motion sensors, biometric sensors, and more. Miniaturization, improved accuracy, and lower power consumption are just a few of the ways that these sensor technologies have progressed. The study also highlights the challenges of integrating AIoT with smart sensing technology. The need for efficient power management, interoperability, the complexity of sensor fusion and data integration, and concerns over data security and privacy are just some of the obstacles in the way.

1. INTRODUCTION

AIoT represents the convergence of Artificial Intelligence (AI) and the Internet of Things (IoT). This integration marries AI's advanced analytical capabilities with the vast interconnected framework of IoT, resulting in intelligent systems capable of autonomous thinking, learning, and decision-making. By amplifying the functions of IoT with the prowess of AI, AIoT seeks to streamline IoT processes, refine interactions between humans and machines, and bolster data handling and analysis. As technology has developed, humans have reached a new age when intelligence and connection pervade every part of our everyday lives. The Internet of Things (IoT) is a paradigm that connects the physical and virtual worlds, ushering in a new age in which everyday items take an active part in online communities. Integrating smart sensing technologies within IoT devices emerges as a critical advancement as we navigate this era of innovation, enabling previously unimaginable data-driven insights, automation, and efficiency (Ramesha & Jeevan. 2020).

The term "Internet of Things" (IoT) refers to the global network of electronic devices, sensors, actuators, and systems that share data and work together to complete tasks. These gadgets cover a wide range of uses, from household gadgets to factory gear to public transportation systems. These Internet of Things devices are distinguished by their sensing, collecting, and transmitting capabilities, which allow for continuous monitoring, analysis, and decision-making in real time.

Significance of Smart Sensing Technologies: Intelligent sensing technologies are the backbone of the Internet of Things' game-changing potential. These developments help bridge the gap between the digital and physical worlds by giving machines the ability to sense and interpret their environments. Sensors such as motion detectors, biometric scanners, image sensors, and environmental sensors give these gadgets a level of perceptual intelligence comparable to that of a human being (G. Naveena Pai, M. Swathi Pai,. 2020). By incorporating these sensors, IoT gadgets may keep tabs on their surroundings, alert users to any shifts they may have missed, and contribute to a dynamic and adaptable environment.

From assessing physiological data to powering autonomous driving systems, sensors have found widespread use in a wide variety of contexts. In the present scientific, technical, and technological fields where computer vision is king, they also play a crucial role in carrying out tasks linked to detection and vision. The Internet of Things (IoT) is an exciting new field that makes use of intelligent sensors, revolving around wireless networks and dispersed sensors to gather data in real time and process it in such a way that it yields predetermined results.

The interplay between sensors and AI is a crucial component of Internet of Things (IoT) gadgets since it gives them feelings and brains. The incorporation of AI enables sensors to function as smart sensors, which benefits a wide range of uses, such as environmental monitoring, tracking different environmental factors, weather forecasting, satellite imaging and utilization, applications based on remote sensing, vigilance over hazard events like landslide detection, driving autonomous vehicles, and improving healthcare (Sridhara & Naveen, 2020).

The use of mobile devices for remote and in-person patient monitoring has increased dramatically in hospitals and clinics as well as diagnostic centers in recent years. Indeed, contemporary sensors have become crucial for intelligent functioning across the board in the scientific and academic communities. Remote sensing, environmental monitoring, and human health surveillance are all made smarter by the ubiquitous use of sensors and the incorporation of IoT. Reflecting the pervasive influence of sensors in virtually every facet of modern life, the field of agriculture applications has also embraced the use of various sensor types over the past decade to monitor and manage environmental parameters like temperature, humidity, soil quality, pollution, air purity, water contamination, and radiation.

1.1. Overview of IoT (Internet of Things) and Its Significance in Modern Technology

Overview of IoT (Internet of Things): The term "Internet of Things" (IoT) is used to describe the global network of physical things, gadgets, machines, and other items equipped with sensors, software, and internet access. These gadgets are able to exchange data with one another and with centralized servers without any assistance from a human operator.

Significance in Modern Technology

The overview of IoT Significance in Modern Technology is presented in Figure 1.

Data Collection and Insights: Sensors built into IoT gadgets provide continuous, real-world data collection. Temperature, humidity, light level, motion, and other environmental factors can all be better understood with the use of this information (Haque & Ahmad, 2023). These discoveries help us make better judgments and enhance our operations.

Automation and Efficiency: Automation is a key feature of the Internet of Things. Connected devices can exchange data and perform tasks in response to specific conditions. Smart thermostats, for instance, can optimize heating and cooling loads by taking into account building occupancy and the time of day.

Smart Homes: With the advent of IoT, "smart homes" have undergone a sea change. Lighting, locks, cameras, and home appliances are just some of the smart home gadgets that can be operated from afar with the use of a smartphone or voice assistant. The ease, safety, and efficiency are all improved by this.

Industrial Applications (IIoT): The term "Internet of Things" (IoT) is more commonly heard in consumer contexts than in industrial ones. Predictive maintenance, remote equipment monitoring, supply chain optimization, and process improvement are all possible with its help. The use of IIoT can boost productivity, decrease downtime, and save money (Kauser & Tanguturi, 2023).

Healthcare and Wearables: Fitness trackers and remote patient monitoring systems are just two examples of IoT-enabled medical gadgets and wearables that enable constant health tracking and real-time data sharing with healthcare practitioners. This allows for preventative measures to be taken, which ultimately benefits patients.

Environmental Monitoring: Sensors connected to the Internet of Things can be used to track environmental factors including air and water quality as well as pollution levels. This information is useful for comprehending environmental shifts and determining how best to adapt to them.

Smart Cities: Improvements in trash management, traffic flow, parking, and public transportation are only some of the urban services that benefit from IoT's implementation in smart cities (Kavitha & Kishore Kumar, 2023). This results in less traffic, better use of available resources, and a higher standard of living for all.

Agriculture and Farming: Sensors in IoT-enabled farming systems track environmental factors including temperature and humidity, as well as plant vitality. Improved crop yields, lower water consumption, and less waste are all outcomes of this data-driven strategy.

Figure 1. Overview of significance in modern technology

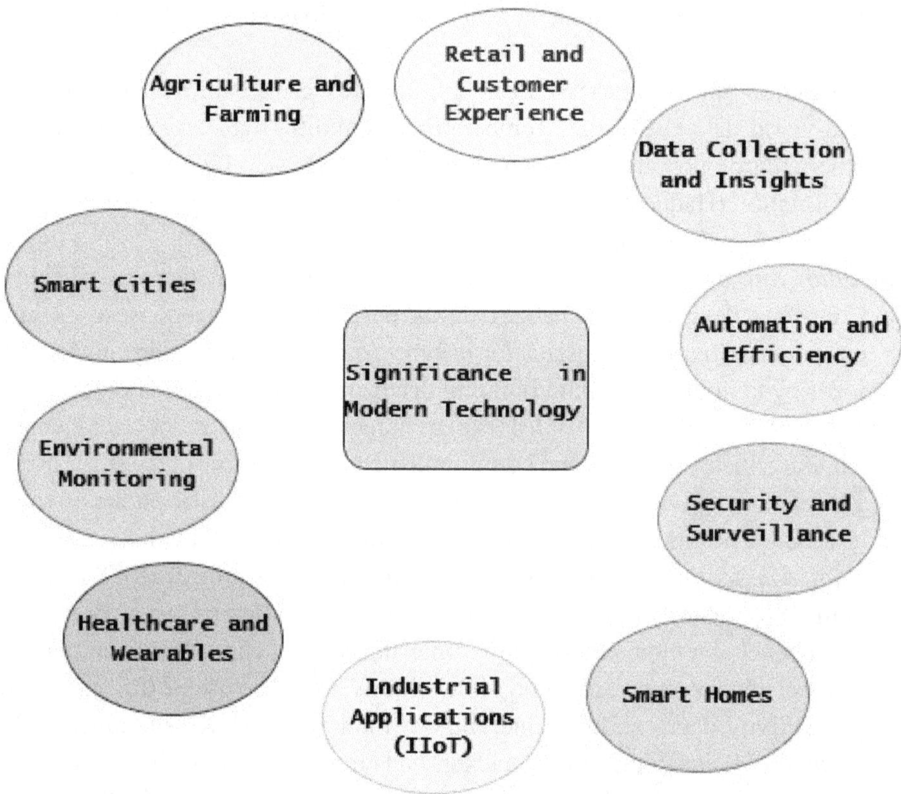

Retail and Customer Experience: The Internet of Things is used by retailers to keep tabs on stock, observe client patterns, and tailor purchases to each individual. Beacons in stores, for instance, can send deals based on the customer's precise location to their mobile devices.

Security and Surveillance: Internet of Things (IoT) security systems use cameras, motion detectors, and access control devices to keep an eye on a building in real time (Kawale & Prasad, 2023). When suspicious activity is detected, alerts can be sent to the owner's smartphone.

The Internet of Things, or IoT, is a network of interconnected computing devices, sensors, and other nodes that enables new forms of business automation and personal insight. Its value comes from the improvements in productivity, comfort, and receptiveness it brings to our surroundings. The benefits of the Internet of

Things (IoT) are undeniable, but its rapid expansion is threatened by issues with data security, interoperability, and ethics.

1.2 Smart Sensing Technologies and Their Role in IoT Devices

The term "smart sensing technologies" is used to describe the practice of incorporating multiple sensors into a device so that it can learn about its surroundings, situations, and user interactions. The purpose of these sensors is to gather data from the physical environment and convert it into digital signals that can be accessed by networked systems like the Internet of Things (IoT). Smart sensors are capable of measuring a broad range of environmental factors, such as pressure, motion, light, and sound, as well as temperature and humidity (M et al., 2023).

Role of Smart Sensing Technologies in IoT Devices

Smart sensing technologies must be included in IoT devices for the IoT ecosystem to operate at its full potential. The following are some benefits provided by sophisticated sensing technologies:

Data Collection: Smart sensors serve as the Internet of Things' "eyes" and "ears". They produce real-time data while continuously observing their surroundings. The subjects covered by this data might range widely, from user behaviors to weather trends.

Environmental Awareness: The data gathered by sensors like those used to measure temperature, humidity, and light levels is invaluable. These sensors can be installed in IoT devices to monitor indoor environments and improve the quality of life, energy efficiency, and plant growth in smart agriculture (N & Farooq, 2022).

User Interaction: Devices can respond to user input thanks to sensors like touch, proximity, and gesture recognition. Examples include smart speakers, which can be operated by voice, and other voice-activated devices and smart screens.

Automation and Decision-Making: By collecting and analyzing data in real time, smart sensors provide IoT devices the ability to act responsibly. A smart thermostat, for instance, can make heating and cooling adjustments in response to temperature readings, so maximizing efficiency and occupant comfort (Ashreetha & Anandaram, 2023).

Safety and Security: Traditional examples include motion sensors and cameras used in security systems. Improve safety and prevent would-be intruders with the use of motion detectors, cameras, and alarms/notifications.

Healthcare and Wellness: Wearable Internet of Things devices equipped with biometric sensors (like heart rate monitors) let users keep constant tabs on their vital signs. These gadgets have the potential to notify users and medical professionals to important information at just the right time.

Predictive Maintenance: Machine sensors can measure things like vibration and temperature for IIoT applications. Sensor anomalies may point to underlying problems, prompting inspection and repair before failure (Kawale & Prasad, 2023).

Environmental Monitoring: The air quality in cities may be monitored by IoT devices equipped with gas sensors. This information is useful for gauging pollution levels, backing up regulatory efforts, and directing public health measures.

Real-Time Analytics: Smart sensors capture data that can either be sent to centralized systems for analysis or evaluated locally (on the device itself) utilizing artificial intelligence and machine learning techniques. This paves the way for immediate analysis and action.

***Optimizing Resource Utilization*:** Smart irrigation systems utilize soil moisture sensors to accurately irrigate plants, cutting water waste and increasing production (Chinamuttevi & Prasad, 2022).

Smart sensing technologies are the foundation upon which IoT devices are built, allowing them to engage with the real world, respond to environmental changes, and deliver actionable data-driven insights. The convergence of these innovations with the Internet of Things creates new avenues for technological advancement across all sectors, raising standards of productivity, comfort, and well-being.

1.3 Importance of Integrating Smart Sensing Technologies Into IoT for Enhanced Functionality and Data-Driven Insights

There are several ways in which functionality can be improved and data-driven insights can be gained through the incorporation of smart sensing technologies into the Internet of Things (IoT). The significance of this amalgamation can be shown in the following examples:

Enhanced Functionality

Diverse Data Acquisition: IoT devices may now gather a wide variety of data from the physical environment thanks to smart sensing technology. Conditions in the surrounding area (temperature, humidity, light levels, etc.), human input, motion, and other factors are all part of this information. When these technologies are combined, the capabilities of IoT devices grow, allowing them to function in additional situations (Mali & Hemant, 2022).

Real-Time Monitoring: Smart sensors allow Internet of Things gadgets to constantly monitor their environments and react instantly to changes. For instance, if motion is detected by a smart home security system, the owners will receive warnings right away. This instantaneous response is a major boon to security.

Contextual Awareness: Intelligent sensors give IoT gadgets a sense of perspective. Technology can make good choices by seeing and learning from its surroundings. A smart lighting system, for instance, can modify illumination in response to environmental light conditions, so maximizing efficiency and occupant satisfaction (P et al., 2022).

Data-Driven Insights

Actionable Analytics: Data streams generated by smart sensing technology can be evaluated for insights. Businesses, people, and groups can all benefit from having a better grasp of trends and patterns so that they can make more educated choices.

Predictive Capabilities: Smart sensors collect information that can be used for forecasting purposes. Predictive maintenance, for instance, makes use of sensor data to foresee equipment failures, allowing servicing to be scheduled in advance of any impending breakdowns.

Optimization of Processes: Processes can be improved with the use of data-driven insights. In farming, for instance, sensors can track the amount of moisture in the soil, allowing for more efficient irrigation and higher agricultural harvests (Sharma & Rao, 2022).

Personalization: Intelligent sensing technologies aid in the creation of unique and engaging encounters. Wearable gadgets with biometric sensors can monitor health indicators and give consumers individualized guidance on how to best take care of themselves.

Early Detection: Internet of Things gadgets with high-tech sensors can spot unusual behavior. This kind of early detection has applications ranging from fixing broken machinery to monitoring environmental changes.

Energy Efficiency and Resource Conservation

Intelligent Control: Devices with smart sensors can respond accurately to their environments. A smart thermostat, for instance, can save money on utilities by adjusting the temperature based on whether or not the building is occupied.

Optimal Resource Allocation: Smart cities use sensors to track vehicle movements and automatically modify traffic lights to ease congestion, increase efficiency, and cut down on emissions and fuel costs.

Water and Energy Management: Smart sensors provide for more efficient control of utilities like water and energy in commercial and domestic settings. One way to save money and water is to install leak detecting sensors.

Real-Time Decision-Making

Edge Computing: Connecting smart sensors to IoT gadgets enables edge computing, in which information is processed close to its point of collection rather than being transmitted to a remote server. This allows for quicker judgments to be made, which in turn allows for instantaneous responses.

Faster Responses: In order for equipment to quickly adapt to new conditions, real-time data from smart sensors is essential. For instance, a traffic control system that makes use of the Internet of Things can quickly adjust to new circumstances (Swetha & Namitha, 2022).

In summary, greater functionality and data-driven insights can be unlocked through the incorporation of smart sensing technologies into IoT devices. Incorporating these innovations into everyday life and business creates the potential for smarter, more adaptive systems in all spheres of activity.

2. ARTIFICIAL INTELLIGENCE OF THINGS (AIOT)

Artificial intelligence of things (AIoT) is a potent new hybrid formed by the coming together of rapidly developing technologies like AI, IoT sensors, and 5G networks. Although the Internet of Things (IoT) is still in its infancy, it is already at the helm of a revolutionary wave, shaping a wide range of applications and trends that have the potential to radically alter the landscape of commercial opportunities. More generally, the Internet of Things system allows for unhindered data flow, which may subsequently be used in tandem with artificial intelligence methods to produce useful insights. By combining these two processes, not only is the data consolidated, but

Figure 2. Integration of IoT to AI technique

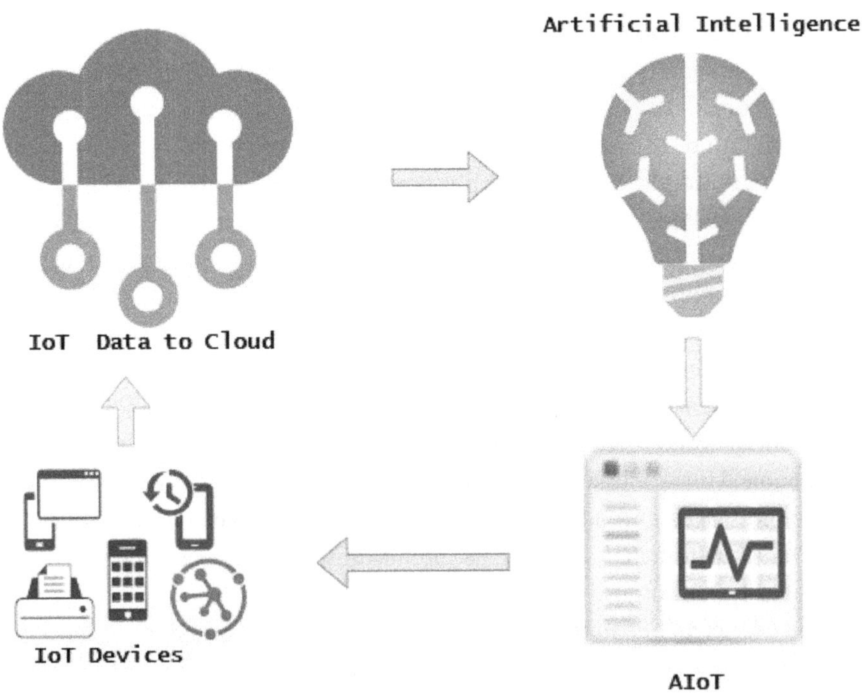

it can also be interpreted automatically, opening the door to applications like image analysis and predictive data processing. Figure 2 graphically depicts the interplay between AI and IoT, showing how this collaboration improves the quality of life for end users.

Managing harvests, greenhouse conditions, and establishing intelligent fertigation procedures that adapt to changes in external factors are some of the primary applications of AIoT technology in agricultural environments. Convolutional neural networks, for instance, can use IoT data to detect and prevent disease in a wide variety of settings and applications (Annepu & Ramesha, 2021). Agriculture operations and management are made more easily accessible, more efficient, and more capable of autonomous functioning by using the power of AIoT. Without a shadow of a doubt, the Internet of Things (IoT) will become a primary catalyst pushing advances in smart agriculture. But there are still challenges to be met, such as the preparedness of embracing technology as a normal practice and the associated costs.

3. SMART SENSING TECHNOLOGIES

Environmental Sensors (Temperature, Humidity, Light, etc.): The purpose of environmental sensors is to keep track of a wide range of environmental conditions. In order to regulate the environment and maximize comfort, temperature sensors are needed. Humidity sensors are crucial for ensuring optimum conditions in places like greenhouses by measuring relative humidity in the air. Photodetectors (also called as light sensors) measure how bright an object is in the surrounding environment, allowing electronics to optimize their illumination output in response to ambient light levels. Smart sensing technologies related to the Internet of Things are depicted in Figure 3.

Motion and Presence Sensors: Movement inside a preset area is detected by the motion sensors. Passive infrared (PIR) sensors are commonly utilized in security systems because of their ability to read heat signatures and locate people. Ultrasonic sensors are able to detect movement or presence by emitting sound waves and analyzing the reflected waves. These sensors are used in various automated systems, including those that regulate lights and provide security for homes.

Gas and Chemical Sensors: Certain gases can be detected by using gas sensors. They play an important role in applications related to safety, such as the detection of carbon monoxide, methane, and other potentially dangerous gases (Mudholkar

Figure 3. Smart sensing technologies

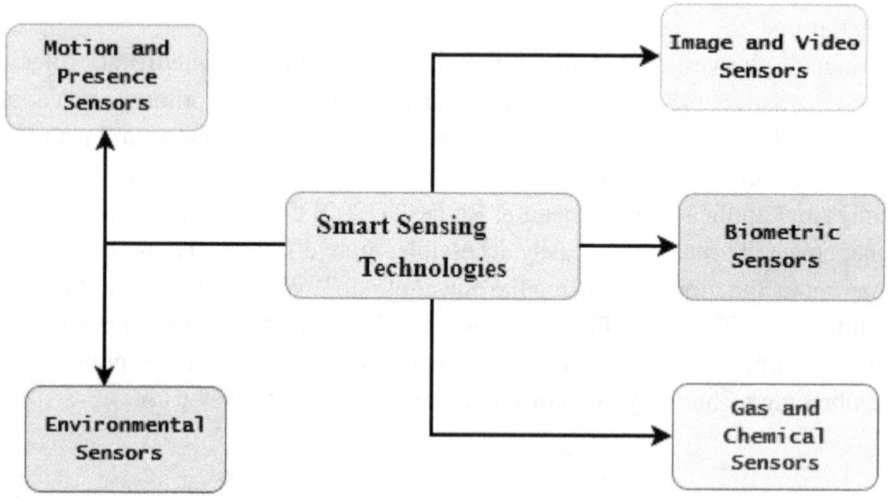

& Mudholkar, 2021). There are a wide variety of uses for chemical sensors, from monitoring the environment to ensuring worker safety in the workplace.

Biometric Sensors (Heart Rate, Fingerprint, etc.): Biometric sensors record information that is both personal and distinctly physiological or behavioral. Heart rate monitors are useful for keeping tabs on your health and fitness since they measure your heart rate using either an optical or electrical method. Improve the safety of your smartphone or building's entry system with fingerprint sensors that use each user's fingerprint to verify their identity.

Image and Video Sensors: Image sensors collect information about what the human eye sees. They take in light and transform it into electronic impulses that can be used to make a picture or a movie. Surveillance systems, photography, and even facial recognition software all rely heavily on image sensors. Object detection and scene analysis are made possible by IoT devices with built-in image recognition capabilities.

Advancements in Sensor Technology

Miniaturization: As microfabrication methods have improved, smaller and more compact sensors have been available. This reduction in size makes it possible to include sensors into previously impractical products like wearables and tiny IoT nodes.

Energy Efficiency: In order to extend the runtime of battery-powered gadgets, modern sensors are made to use as little power as possible. This is especially crucial for Internet of Things gadgets that must run without human intervention over long stretches of time.

Accuracy and Precision: Advances in sensor technology have allowed for more precise data collection. This guarantees the accuracy of the data for crucial uses like medical diagnostics and factory control.

Wireless Connectivity: Wi-Fi, Bluetooth, Zigbee, and other wireless protocols are used to send data from many types of smart sensors. Because of this, connecting to IoT networks is easier than ever.

Real-World Applications

Smart sensors find applications across a wide range of industries and use cases:

Healthcare: Wearables with built-in biometric sensors monitor vitals like heart rate and sleep habits. Sensors can remotely monitor patients and relay real-time data to doctors.

Smart Homes: Sensing devices in the environment allow for more efficient temperature and humidity control.

Industrial IoT (IIoT): Predictive maintenance and reduced downtime are made possible by sensors installed in production facilities.

Environmental Monitoring: In cities, gas sensors monitor the air quality, and in bodies of water, they detect the amounts of pollution.

Automotive: Sensors like LiDAR and radar are used by ADAS to improve vehicle safety and automation.

Agriculture: In order to maximize crop yields while minimizing water waste, farmers are increasingly turning to soil moisture monitors.

Retail: In-store presence sensors can activate digital displays or communicate with customers' mobile devices to create unique shopping environments.

Connected devices may learn about and respond to their surroundings thanks to smart sensing technology. They improve automation, make real-time data collecting possible, and give businesses and consumers more access to reliable data on which to base choices.

4. IOT INTEGRATION CHALLENGES

To ensure the smooth operation and success of IoT systems, numerous obstacles posed by IoT integration must be overcome. To ensure interoperability, quick data sharing, and seamless integration among devices and systems, the IoT ecosystem relies heavily on the standardization of communication protocols. However, IoT's diverse device, application, and communication needs can cause standardization problems. Figure 4 depicts several of the most pressing problems we face.

Interoperability and Standardization: There is a wide variety of IoT devices on the market, and these devices may not all use the same interface, data format, or communication protocol. Devices from different manufacturers may be challenging to integrate due to their inability to communicate with one another. If we want our devices to be able to talk to one another and cooperate without any hiccups, we need to move toward standardization.

Security and Privacy: Because of the sensitive information that IoT devices collect and communicate, they are vulnerable to cyber attacks. It's vital to ensure the safety of all data during transmission, storage, and device authentication. Data

Figure 4. IoT integration challenges

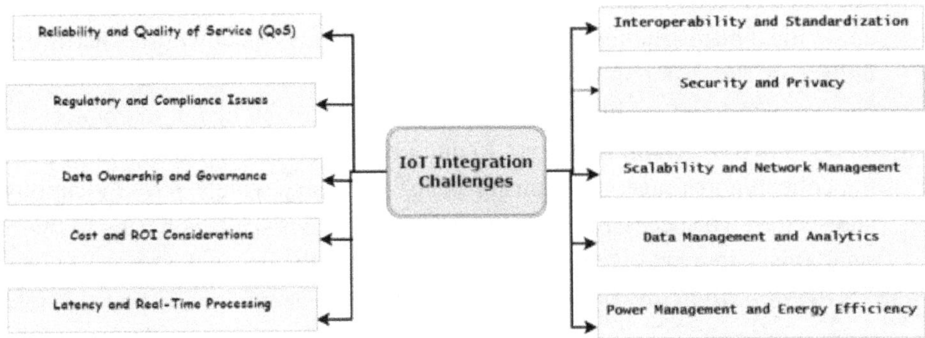

leaks, unauthorized access, and even the compromising of vital infrastructure can result from insufficient security measures. When collecting information about an individual, considerations concerning data use, sharing, and security often emerge.

Scalability and Network Management: The increasing number of connected devices poses new challenges for system administrators. Congestion in the network, decreased efficiency, and higher operating expenses can all result from a lack of scalability. The complexity of an ever-expanding IoT environment calls for network management tools and tactics.

Data Management and Analytics: Extensive volumes of data are produced by the Internet of Things, frequently in real time. It's difficult to find an effective way to store, handle, and analyze all this information. Big data analytics and cloud computing technologies may be necessary to process the massive amounts of data generated by the Internet of Things.

Power Management and Energy Efficiency: There are a lot of IoT devices that run on batteries or have restricted power supplies. To keep devices running longer without requiring as much upkeep or replacement of batteries, effective power management is crucial. This problem can be alleviated in part by employing low-power communication protocols and energy collecting systems.

Latency and Real-Time Processing: Real-time decision making in the context of certain Internet of Things applications, such as autonomous vehicles or industrial automation, necessitates extremely low latency. Edge computing solutions that handle data closer to the source are typically necessary in such circumstances to ensure that data is analyzed and acted upon rapidly.

Reliability and Quality of Service (QoS): High levels of stability and quality of service are essential for IoT applications in fields such as healthcare, transportation, and critical infrastructure. The repercussions of network downtime, hardware failure, or sluggish communication are not to be taken lightly. Redundancy, backup plans, and quality control procedures should all be put into place.

Regulatory and Compliance Issues: Regulations and compliance standards for the collecting, storage, and transfer of IoT data vary widely across locations and sectors. To successfully deploy IoT technologies while adhering to these standards, one must give serious thought to both the law and ethics.

Data Ownership and Governance: It's not always clear who owns and controls data collected by IoT devices. Device makers, end users, and service providers could all have a right to access the information. It is critical to define who owns what in terms of data.

Cost and ROI Considerations: There are a variety of expenses associated with putting into action IoT technologies. Making sure the investment in the Internet of Things (IoT) pays off requires considerable forethought and analysis of the potential gains.

Addressing these problems calls for a comprehensive strategy that integrates stakeholder input, new technologies, regulatory structures, and established best practices. In order to fully realize the promise of IoT and reap its benefits across sectors and applications, these obstacles must be overcome. Here are several practical illustrations from the real world.

Predictive Maintenance: Through AIoT, manufacturing facilities can anticipate machinery breakdowns or maintenance needs. Take Siemens, an international industrial leader, for example. They employ AIoT to track their equipment's degradation. By forecasting service intervals, they minimize disruptions and boost productivity.

Smart Agriculture: John Deere, the US-based agricultural machinery producer, integrates AIoT into its tractors and harvesters. These devices can now evaluate live data concerning soil quality and plant vitality, modifying strategies instantaneously to enhance crop output.

Network Security: As the number of interconnected gadgets skyrockets, security issues persist. The notorious 2016 Mirai botnet assault exploited weak IoT devices, causing substantial internet disruptions. This event underscores the risks associated with insufficient protection of AIoT equipment.

Data Overload: Barcelona, in its quest to become a smart city, amasses extensive data from sensors scattered throughout the metropolis. However, extracting valuable information from this data pool can be daunting without advanced AI techniques and approaches.

5. ADVANCEMENTS IN IOT INTEGRATION

Integration of the Internet of Things has progressed thanks to technological breakthroughs and shifting business requirements. These developments allow for improved functionality, data processing efficiency, and connectivity in IoT systems. Key developments in integrating the internet of things include the following:

Edge Computing and Fog Computing: Edge computing is the practice of processing data at or near its source, rather than sending it off to distant cloud servers. Fog computing is an extension of this principle that allows for real-time data processing and decision-making by bringing distributed computing closer to the edge. These advancements enable quicker answers, reduced network traffic, and generally more effective systems.

5G Connectivity: 5G networks enable more devices to connect to a network simultaneously and faster data transmission. This is particularly useful for Internet of Things (IoT) applications that rely on continuous two-way communication, such autonomous cars and industrial automation. With 5G, more connections may be active at once without affecting performance.

AI and Machine Learning Integration: To process and analyze the vast amounts of data generated by IoT devices, more and more AI and machine learning algorithms are being included into IoT systems. Through this connectivity, gadgets can learn to reason, forecast, and draw useful conclusions from data patterns.

Digital Twins: A digital twin is a digital copy of a real-world product, process, or system. It is possible to perform predictive maintenance, run simulations, and enhance efficiency by linking IoT devices to their digital twins and comparing the resulting data.

Blockchain for Security and Trust: Security, data integrity, and trust in transactions are all improved when blockchain technology is applied to the IoT. Transactions can be recorded in a public and unchangeable ledger, making it hard to forge records. This is especially helpful for managing supply chains and distributing information across multiple stakeholders.

Standardization and Interoperability: There is ongoing work to standardize communication protocols and protocols (such as MQTT, CoAP, and OPC UA). Devices from different manufacturers can now communicate with one another thanks to these standardized protocols.

Data Analytics at the Edge: By processing and analyzing data at the edge, closer to where it is created, less data needs to be sent to central servers. This reduces waiting time and saves bandwidth without sacrificing analytical power.

Advanced Security Solutions: The importance of security in IoT infrastructures grows as their deployment expands. Recent developments in cryptography, authentication, and intrusion detection systems have contributed to the improvement of security for Internet of Things devices and networks.

Integration of Augmented Reality (AR) and Virtual Reality (VR): When Internet of Things (IoT) is combined with AR and VR, the result is a really immersive experience. Technicians, for instance, can use augmented reality glasses to view live data and superimpose that data on top of images of actual machinery.

Hybrid Cloud Deployments: A hybrid cloud utilizes both public and private clouds in addition to traditional in-house resources. Because of the adaptability and scalability of this method, businesses can implement Internet of Things solutions tailored to their own requirements.

Energy Harvesting Solutions: IoT devices may now self-power from renewable sources like solar energy, vibration, or temperature gradients thanks to developments in energy harvesting technologies. This lessens the need for conventional power sources and increases the runtime of portable electronics.

As a result of these developments, IoT systems may now serve a wider variety of applications in fields as diverse as manufacturing, healthcare, transportation, and smart cities than ever before. Innovative methods and techniques will likely continue to be of use in IoT integration as technology develops further.

6. CASE STUDIES

Here are some real-world examples that illustrate the usefulness and advantages of IoT integration:

Smart Agriculture – Precision Farming: Integrating IoT into farming has led to more precise methods of crop production. Fields can be monitored for things like wetness, temperature, and humidity with the help of sensors. This information is sent to a centralized hub where it is processed to ascertain the most effective times to water plants. This method ensures plants get the proper amount of water at the right time, which saves water, cuts costs, and boosts agricultural yields.

Healthcare – Remote Patient Monitoring: Wearables with Internet of Things capabilities, such as smartwatches and fitness trackers, can be used for out-of-office medical monitoring. Heart rate, blood pressure, and glucose levels are just some of the vital indicators that can be continuously monitored and reported to healthcare specialists for patients with chronic diseases. Timely actions are made possible by this real-time data, and the need for frequent clinic visits is diminished as a result.

Manufacturing – Predictive Maintenance: Maintenance procedures in manufacturing have been revolutionized by the Internet of Things. Machines with inbuilt sensors may track data like vibration levels, temperatures, and power consumption. Manufacturers can arrange preventative maintenance on their equipment based on the results of this analysis. By taking this measure, we can keep production interruptions to a minimum and save money.

Smart Cities – Traffic Management: Traffic in smart cities with integrated IoT is more efficient and has less congestion. Traffic data is gathered via sensors on roads and at crossings. This information is utilized to enhance the timing of traffic signals and reroute vehicles in real time. The end effect is less traffic, faster commutes, and fewer greenhouse gas emissions.

Energy Management – Smart Grids: The Internet of Things (IoT) is incorporated into smart grids to control and improve power distribution. Energy usage, voltage levels, and possible defects are tracked via sensors embedded in power lines and substations. Control centers receive this information and use it to better manage energy consumption, prevent power outages, and balance supply and demand.

Retail – Inventory Management: Internet of Things devices are used by stores for accurate stock counts in real time. Products tagged with Radio-Frequency Identification (RFID) chips can have their stock levels automatically monitored. The system tracks sales and orders more of a product when inventory drops below a certain level. The likelihood of shortages and surpluses is reduced.

Environmental Monitoring – Air Quality: Cities use air quality monitoring stations equipped with Internet of Things sensors to track airborne pollutants in real time. Particulate matter and nitrogen dioxide are two examples of pollutants that can be detected by sensors. To better inform policy choices, alert citizens to bad air quality, and monitor progress in lowering pollution, the data is made public and shared with local authorities.

Logistics – Supply Chain Visibility: By allowing for constant monitoring of inventory locations and conditions, IoT integration improves supply chain management. Shipments can have their position, temperature, humidity, and vibrations monitored using sensors. Companies can use this information to improve their routes, monitor product quality, and react to unexpected events more effectively.

These examples show the wide variety of domains where the Internet of Things is having a profound effect. Increased productivity, better data for better decisions, and more satisfying interactions with products and services are all possible thanks to the Internet of Things.

7. ETHICAL IMPLICATIONS

In the swiftly changing realm of AIoT, where advanced technologies seamlessly integrate into our daily existence, the potential for progress and innovation is exhilarating. However, along with these technological strides come intricate ethical quandaries that demand our consideration. As we infuse intelligence into items, spaces, and entire cities, distinctions between ease and invasion, freedom and oversight, often become indistinct. The growing prevalence of interconnected gadgets heightens issues of privacy and observation, but it also unveils subtle conundrums like inherent biases in algorithmic choices and the broader societal repercussions of pervasive automation. Engaging with these ethical facets is more than just a theoretical endeavor; it's a crucial journey, ensuring the intelligent world we craft places human respect, freedoms, and collective ethics at its heart.

Privacy Concerns: As smart devices become more prevalent, surveillance concerns rise. Take, for instance, Amazon's Ring doorbell, which drew scrutiny for providing video footage to law enforcement agencies without the approval of the homeowners.

Bias in Algorithms: AI systems can inadvertently adopt prejudices present in their training data. When AIoT devices use such skewed algorithms, they risk magnifying these inherent biases. A case in point is facial recognition in certain intelligent monitoring systems, which has shown disparities in accuracy for different ethnicities, potentially resulting in unfair outcomes.

Dependence on Technology: As our homes, urban areas, and industrial sectors grow increasingly intelligent, our dependence on technology intensifies. This heightened reliance can introduce susceptibilities, where system malfunctions or cyber-attacks might lead to severe repercussions.

Job Displacement: The efficiency brought about by AIoT can result in the automation of roles traditionally performed by people. This enhances productivity, but it also raises worries about potential job displacement. For example, as warehouses integrate AIoT into their inventory systems, numerous manual positions could become obsolete.

Standing at the juncture of an AIoT-influenced horizon, it's vital to proceed with both prudence and vision. While the immense promise of this merger is undeniable, it shouldn't eclipse the moral duties intertwined with it. Welcoming the benefits and novelties of AIoT also necessitates upholding the tenets of equity, clarity, and regard for personal liberties. A genuinely 'intelligent' society isn't defined merely by its technological might but by its capacity to utilize such innovations for communal welfare, ensuring they aid humanity without undermining the distinct attributes that define human essence.

8. FUTURE TRENDS

Here are some future trends in the field of IoT integration:

Edge AI and Processing: Since IoT devices produce vast quantities of data, it is becoming increasingly vital to handle this data locally, or at the edge. Edge AI is the practice of implementing ML and AI algorithms locally, on edge nodes or IoT devices. This allows for more efficient use of network resources, quicker decision-making, and lower latency.

5G and Beyond: When 5G networks become available, IoT connection will be completely transformed. More real-time applications like driverless vehicles, remote surgery, and immersive virtual experiences will be possible thanks to its fast data speeds, ultra-low latency, and capacity to link a large number of devices simultaneously. In order to improve the IoT even more, research into technologies such as 6G is already underway.

Blockchain and IoT Security: Increased security, data integrity, and trust can be achieved through the integration of blockchain technology into IoT devices. There is less chance of data tampering or illegal access with blockchain's secure and transparent transaction records. This is especially helpful in supply chain management, healthcare, and the financial sector.

Swarm Intelligence: An example of a natural system that swarm intelligence mimics is an ant colony or a flock of birds. Using this idea, IoT devices can work together independently to address challenging issues. Possible uses include managing traffic flow in smart cities and self-organizing sensor networks for use in emergency situations.

AI-Driven Predictive Analytics: Predictive analytics for the Internet of Things will benefit greatly from the use of AI and machine learning. The ability of AI systems to forecast outcomes, optimize processes, and spot anomalies will improve as they learn from both historical and real-time data. For use cases like predictive maintenance and tailoring the user experience, this is invaluable.

Energy Harvesting and Sustainability: Energy-efficient Sustainability relies heavily on Internet of Things devices. As energy harvesting technology improves, more gadgets will be able to run on clean, renewable power like the sun, motion, or heat. IoT installations will have less of an effect on the environment, and battery life will be increased.

Digital Twins Evolution: There will be an expansion of the capabilities of digital twins, which generate digital copies of physical items, to include more intricate structures. From manufacturing to city planning, digital twins will play a crucial role in IoT integration by facilitating real-time simulation, optimization, and scenario analysis.

Hybrid Cloud Architectures: Hybrid cloud designs, which integrate public cloud, private cloud, and on-premises resources, will become increasingly popular for Internet of Things installations. With this method, you can optimize the price and performance of your IoT applications without sacrificing adaptability, scalability, or control over your data.

Human-Machine Convergence: Convergence between humans and machines will increase as IoT, AI, and wearables all work together. New methods of interaction between humans and technology, such as brain-computer interfaces, biometric sensors, and augmented reality, will make for more intuitive and immersive digital experiences.

Regulatory and Ethical Considerations: Policymakers and regulators will address concerns about data privacy, security, and ethical usage of IoT technologies as their prevalence increases with IoT integration. Trust among users is essential to the success of the Internet of Things, thus regulations and guidelines will develop to ensure that IoT deployments meet ethical and legal requirements.

These developments are evidence of the ever-increasing pace at which IoT integration is advancing in response to the increasing need for interconnected, intelligent, and efficient systems in all sectors of the economy.

9. CONCLUSION

In conclusion, the widespread adoption of IoT technology has dramatically altered the way we engage with the physical world. Intelligent ecosystems have emerged as a result of the convergence of devices, data, and connectivity, leading to increased productivity, better decision-making, and more satisfying user experiences in a wide range of fields and disciplines. Smart homes, which improve energy efficiency and security, and industrial IoT systems, which allow for predictive maintenance and faster operations, are just two examples of the many positive outcomes of IoT integration. Smart sensing technology developments have given devices the ability to see and comprehend their surroundings, while standardization initiatives have made it possible for devices from different manufacturers to communicate with one another in a smooth manner. The future of IoT integration is uncertain because the field is changing so quickly. The potentials of the Internet of Things are evolving in response to developments like edge computing, 5G connection, AI-driven analytics, and eco-friendly efforts. Not only are industries benefiting from these developments, but so are cities, healthcare, and environmentally friendly methods of operation.

However, obstacles such as establishing strong security, responding to privacy concerns, and understanding regulatory landscapes persist. The success of IoT integration depends on both technological innovation and responsible execution that takes into account ethical concerns and protects users' personal information. Industry players, researchers, legislators, and the larger community must work together to adapt to this ever-changing environment. Standards, best practices, and innovative solutions that maximize the benefits of IoT integration while limiting possible hazards will be developed through collaboration as IoT continues to expand its reach.

IoT integration is a game-changer because it allows us to create a world that is more interconnected, intelligent, and reactive. We can influence the future in which data-driven insights and smart interactions improve our daily lives, reshape entire industries, and help create a more sustainable and linked global society by embracing the possibilities of IoT technologies.

ACKNOWLEDGMENT

We would like to acknowledge the support given by B.M.S. Institute of Technology and Management, Bangalore, Karnataka, India.

REFERENCES

Arudra Annepu, M. (2021). IoT Enabled Smart Lighting System for Smart Cities. *Journal of Physics: Conference Series*, *2089*(1), 012037. doi:10.1088/1742-6596/2089/1/012037

Ashreetha, Anandaram, Gupta, & Verma. (2023). IoT Wearable Breast Temperature Assessment System. *2023 7th International Conference on Computing Methodologies and Communication (ICCMC)*, 1236-1241. 10.1109/ICCMC56507.2023.10083511

Avinash Sharma, Rao, Shankar, Sarma, Chaturvedi, & Hussain. (2022). Industrial quality healthcare services using Internet of Things and fog computing approach. *Measurement: Sensors, 24*. doi:10.1016/j.measen.2022.100517

Chinamuttevi, V. S., Prasad, K., Begum, A. Y., Tanguturi, R. C., & Ahmed, B. (2022). Implementation of an Early Warning System using Internet of Things and Rainfall Threshold. *2022 6th International Conference on Electronics, Communication and Aerospace Technology*, 491-496. 10.1109/ICECA55336.2022.10009211

Haque, M. A., Ahmad, S., Eljialy, A. E. M., Uddin, M. Y., & Sonal, D. (2023). Internet of Things (IoT) based Model for Water Management System. *2023 International Conference on Smart Computing and Application (ICSCA)*, 1-5. 10.1109/ICSCA57840.2023.10087587

Kauser, S. H., & Tanguturi, R. C. (2023). Implementation of Machine Learning Approach for Detecting Cardiovascular Diseases. *2023 3rd International Conference on Intelligent Technologies (CONIT)*, 1-6. 10.1109/CONIT59222.2023.10205893

Kavitha, R., Kishore Kumar, R., Pandidurai, M., & Senthil Kamalesh, M. S. C. (2023). Design of IoT based Rural Health Helper using Natural Language Processing. *2023 4th International Conference on Electronics and Sustainable Communication Systems (ICESC)*, 328-333. 10.1109/ICESC57686.2023.10193519

Kawale, S. R., Prasad, K., Anil Kumar, N., Reddy, N. S., & Ashreetha, B. (2023). Technologies for Comprehensive Information Security in the IoT. *2023 International Conference for Advancement in Technology (ICONAT)*, 1-5. 10.1109/ICONAT57137.2023.10080332

Kawale, S. R., Prasad, K., Palanikkumar, D., Mary, P. A., & Begum, A. Y. (2023). A Novel IoT Framework and Device Architecture for Efficient Smart city Implementation. *2023 7th International Conference on Trends in Electronics and Informatics (ICOEI)*, 420-426. 10.1109/ICOEI56765.2023.10125677

Mali, Tirmare, Suryawanshi, & Chaturvedi. (2022). Novel Predictive Control and Monitoring System based on IoT for Evaluating Industrial Safety Measures. *IJEER, 10*(4), 1050-1057. doi:10.37391/ijeer.100448

Mudholkar, P., Mudholkar, M., Puneeth Kumar, B. S., Dankan Gowda, V., & Raju, S. S. (2021). Smart Villages: IoT Technology Based Transformation. *Journal of Physics: Conference Series, 2070*(1), 012128. doi:10.1088/1742-6596/2070/1/012128

N., Farooq, M, Reddy, & Tanguturi. (2022). Improved Secure Communication Mechanism for IoT Platform Devices. *2022 International Conference on Smart Generation Computing, Communication and Networking (SMART GENCON)*, 1-6. 10.1109/SMARTGENCON56628.2022.10083714

Naveena Pai, Swathi Pai, Shruthi, & Naveen. (2020). Internet of Things: A Survey on Devices, Ecosystem, Components and Communication Protocols. *2020 4th International Conference on Electronics, Communication and Aerospace Technology (ICECA)*, 611-616. . doi:10.1109/ICECA49313.2020.9297458

R. M., Nareshkumar, & Gururaj. (2023). An Integrated IoT Technology for Health and Traffic Monitoring System with Smart Ambulance. *2022 IEEE North Karnataka Subsection Flagship International Conference (NKCon)*, 1-6. 10.1109/NKCon56289.2022.10126786

Ramesha, M., Jeevan, K., & Sathisha, B. M. (2020). Implementation of IoT Based Wireless Electronic Stethoscope. *2020 Third International Conference on Multimedia Processing, Communication & Information Technology (MPCIT)*, 103-106. 10.1109/MPCIT51588.2020.9350476

Sharma, P. A., Reddy, S., & Patwal, P. P. S. (2022). Data Analytics and Cloud-Based Platform for Internet of Things Applications in Smart Cities. *2022 International Conference on Industry 4.0 Technology (I4Tech)*, 1-6. 10.1109/I4Tech55392.2022.9952780

Sridhara, S. B., Naveen, K. B., Ramesha, M., & Pai, G. N. (2020). Internet of things: Internet revolution, impact, technology road map and features. *Adv. Math. Sci. J.*, *9*(7), 4405–4414. doi:10.37418/amsj.9.7.11

Swetha, Namitha, Manu, Rashmi, & Chinamuttevi. (2022). IOT Based Smart Health Care System to Monitor Covid-19 Patients. *IJEER*, *10*(1), 36-40. doi:10.37391/IJEER.100105

Chapter 4
An Overview of Blockchain Applications and Benefits in Cloud Computing

Ramiz Salama
Near East University, Nicosia, Turkey

Fadi Al-Turjman
AI and Robotics Institute, Near East University, Nicosia, Turkey & Faculty of Engineering, University of Kyrenia, Kyrenia, Turkey

ABSTRACT

Due to its many attributes, including decentralization, transparency, and security, blockchain is now a very important and novel technology for the current generation of industrial applications. One of these fields, the cloud of things, was created by fusing cloud computing and the internet of things. Blockchain is a key component of this technology because it provides data privacy and network security, eliminates cloud of things challenges through decentralization, and incorporates cloud of things features through elasticity and scalability functionalities, enhancing the overall efficiency of blockchain operations. As a result, it is highly advantageous to connect blockchain to the internet of things (IoT); this technology is known as BCoT, and it is now thought to be a very promising field for a wide range of industrial applications.

INTRODUCTION

A blockchain is a continuously growing linked list that resembles a record's structure. It consists of blocks joined by connections, and the process of storing information

DOI: 10.4018/979-8-3693-0786-1.ch004

in blocks is encrypted. Every block on the blockchain has transaction information, a timestamp indicating when the block was added to the chain, and a cryptographic hash of the block before it. The preceding block cannot be deleted or altered since each block is connected to and carries data about the one before it. Blockchains can avoid changing any data in any of the blocks because, once recorded, the contents of any block cannot be modified without also changing the contents of all the blocks that came before it. Significant improvements in information technology and communication are being made by both the Internet of Things (IoT) and cloud computing. Living standards have significantly changed as a result of the Internet of Things (IoT), which has also given rise to numerous new commercial, consumer, and industrial service-related applications that have improved and enhanced our quality of life. The Internet of Things (IoT) is a system of physically connected objects that can connect to and exchange data with other devices and systems via the Internet. These connected objects are outfitted with sensors, software, and other technologies. This system has the ability to recognize data in the environment, store it, process it, extract important information from it, and take actions in response to the information (Adeniyi et al., 2022; Ahmed et al., 2022; Habib, Sharma, Ibrahim, Ahmad, Qureshi, & Ishfaq, 2022; Kaur et al., 2022; Khanna et al., 2022). Due to the constrained resources of IoT devices, such as storage capacity, IoT applications are always outsourced to the Cloud of Things (CoT) model. For processing and administering IoT services, the CoT offers a stable and flexible cloud computing environment, which has the potential to significantly improve system performance and efficiency.

AIM OF THE RESEARCH

The goal of this research is to investigate the emerging technology known as Blockchain and learn more about it, then investigate Cloud Computing and learn more about it. The benefits and applications that can be produced by combining these two technologies won't be apparent until then.

THE NEED FOR BLOCKCHAIN AND COT INTEGRATION

The following are some reasons why conventional CoT networks are ineffective: Traditional CoT confronts a basic challenge in that it relies heavily on centralized communication systems and a single cloud, making it challenging to scale service operations when IoT networks grow significantly. The need for most contemporary CoT systems to outsource IoT data processing to any third party, such as a cloud provider, raises concerns about data privacy, which is the second major challenge that CoT faces. The third major barrier is that centralized network infrastructure increases IoT device power consumption and communication latency, making large-scale CoT deployments in practical applications difficult. Given the limitations of CoT and the advantages of Blockchain, fusing the two technologies could potentially address CoT's issues (Habib, Sharma, Ibrahim, Ahmad, Qureshi, & Ishfaq, 2022; Sathiya et al., 2023).

CLOUD COMPUTING'S POTENTIAL USE OF BLOCKCHAIN TECHNOLOGY

Some significant areas where blockchain can outperform traditional cloud solutions include the ones listed below:

The blockchain system may be quite helpful in terms of enhancing functionality or performance because of its unique qualities. Because blockchain is a decentralized system, it can help develop an architecture in which numerous computers can work together to accomplish one specific task, such data processing or storage, at once, speeding up data processing and uploading. Another concern is enhancing cloud computing security (Fateh & Salarnejad, 2022; Patan et al., 2023; Rabbani et al., 2022). Data security risks always exist since cloud computing is centralized. For instance, if the system as a whole were compromised, there would be no backup data, and any lost data would be irretrievably destroyed. In order to address these issues, blockchain technology has the potential to be used in cloud computing.

BLOCKCHAIN IN CLOUD COMPUTING BENEFITS

Decentralization

The reliance on a centralized server for data management and decision-making presents a fundamental barrier for IoT and cloud computing technologies. For instance, if the central server fails, the entire system may be affected, and important data stored on the server may be lost. The main server is also susceptible to hacker attacks. In a decentralized system, the likelihood that the entire system will malfunction if one server dies is decreased when many copies of the same data are kept on other computer nodes (Al-Turjman & Salama, 2021a). Additionally, data loss should be prevented because there are numerous copies of the data stored on different nodes.

Enhanced Data Security

IoT stored data is typically associated with the home owner's personal information, such as video footage, voice recordings, household items, property, and personal habits, and a leak of this data can result in robbery, attacks, and the illegal sale of personal information for money. This makes storing IOT stored data on the cloud a significant challenge. These circumstances lead to a breach of the cloud infrastructure. The use of blockchain in cloud computing has the potential to increase overall system security as a response to this problem.

Improved Product and Service Ownership Tracing

Keeping track of every vehicle in its network, their present location, the amount of time a vehicle spent at a site, and establishing contact between several cars is one of the logistics industry's most challenging issues. Software solutions like shipment tracking systems face issues due to their centralized nature. Blockchain holds tremendous potential for monitoring these commodities and services.

Tolerance for Faults

The replication of blockchain data across a network of computer systems connected safely via collaborative clouds may benefit from the cloud. This makes it less likely that any cloud node loss will result in a single point of failure, enabling continuous services.

Scalability

Large-scale blockchain applications have the potential to significantly increase the number of transactions on blockchain networks. As a result, robust data processing services are needed for scalable blockchain systems with high transaction execution. The scalability of the cloud enables it to offer this market's blockchain activities on-demand processing capacity (Al-Turjman & Salama, 2021b). By merging blockchain with cloud computing technologies, it is possible to create an integrated system that is highly scalable.

Figure 1. Benefits of blockchain on cloud

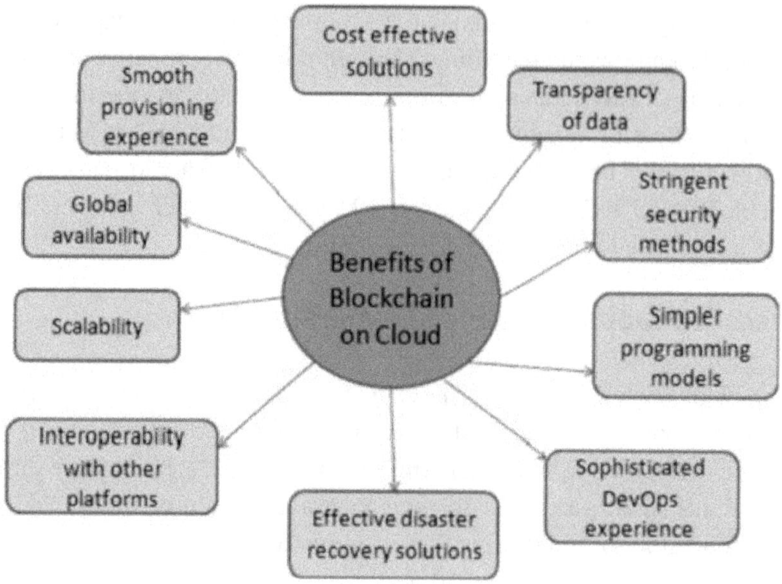

COT AND BLOCKCHAIN INTEGRATION APPLICATIONS

Healthcare Industry Intelligence

For modernizing and improving current healthcare infrastructure and practices, BCoT has great promise. The healthcare sector is made up of medical facilities and organizations. The mainstays of the industry include healthcare-related services, medical devices and technologies including ventilators, and medical insurance. Significant service effectiveness and industrial security challenges can be resolved by BCoT. By reducing the amount of time needed for sharing communication from patients' devices to doctors' devices and among connected devices, new smart services, such as CoT enabled Health data sharing, can be provided with the aid of BCoT integration in healthcare. Electronic health records (EHRs), for instance, will be processed and kept on the cloud, and doctors will have access to patient medical data via mobile devices (such smartphones) for health monitoring (Salama & Al-Turjman, 2022). Blockchain may prove to be especially successful in addressing security issues in the process of exchanging health data because it makes use of decentralized data verification of all peer nodes and message validation through a consensus technique.

Blockchain-Based Smart Home Automation

Smart home automation is one of the most significant industries where BCoT might be applied. A house can become a smart home by automating its appliances, providing comfort for its occupants. A network of Internet of Things (IoT) devices that collect environmental data, store it in the cloud, process it, and then carry out a particular action depending on the processed data constitutes a smart home. When a fire is detected, a message is sent to the homeowner and the house's water sprinkler system or fire alarm is immediately activated. A fire detection system, for instance, comprises of a temperature measuring sensor that determines whether or not there is a fire in the home. Blockchain automation could be applied to smart homes to increase security by lowering the possibility of data loss or privacy invasion. A decentralized data integrity architecture that uses blockchain can be used to guarantee the stability and security of the system without the requirement for a third party.

Integrated Blockchain for Smart Transportation

In the contemporary world, the transportation sector has a lot of space for growth. As sensor, computer, and communication technology have expanded quickly in recent years, so have transportation systems. This has had a significant impact on people's lives, including commute times, road safety, and freight transit times. Smart transportation is an Internet of Things (IoT) application pertaining to included communication technology and vehicle services in transportation networks. However, dynamic vehicle-to-vehicle communication in automotive situations and an over-reliance on centralized network authority have given birth to security flaws (Al-Turjman & Salama, 2020). Blockchain can help create a secure, reliable, and decentralized ITS system in this scenario.

Intelligent Manufacturing

A new industry that could profit from BCoT is smart manufacturing. Smart manufacturing is the use of automated devices that can perform particular tasks far more intelligently than they now can. This sector makes use of service-oriented manufacturing, cloud manufacturing, and IoT-enabled technology (). However, in the existing environment, problems like centralized industrial networks and third-party authority hinder smart manufacturing. The centralized industrial designs lose some of their flexibility, effectiveness, and security. These issues might be resolved by utilizing BCoT, which can improve security while still carrying out the decentralized concept.

ISSUES WITH THE CURRENT CLOUD COMPUTING SYSTEM

According to a Gartner survey of 110 senior executives who commented on the difficulties their various organizations are facing, cloud computing is at the top of the list of rising hazards. How to prevent unauthorized access to a platform with such a huge audience was the main problem that attracted attention. Due to the complexity of cloud computing, fraudsters from all over the world are searching for viruses and system defects that can steal sensitive and personal data. Network failures and a failure to provide access to information may also occur from a cloud provider's inability to function as a result of a disruption. Many free service providers now provide the advantage of storing databases and information. However, some websites take an unreliable position by eliminating content they find offensive without providing customers with sufficient justification. The likelihood of data loss and control dispersion thus rises. An organization's legal requirements might not be satisfied by cloud computing, claims renowned international cryptographer Burce Schneier. The complexity of cloud-based centralized computing and the fast evolution of the regulatory architecture make it more challenging to establish trust and obtain authorization. Although cloud-based businesses have been successful in large firms, small businesses have not been drawn to them because of their high cost and feasibility, which is another depressing trend. The consolidation of cloud services is one of the biggest upheavals, say academics and experts. Data must travel through tightly regulated central servers, which raises costs and reduces efficiency and connectivity. When repositories are centralized, issues with privacy, security, and assessing verifiability arise.

SOLUTION

These servers and repositories may be decentralized in the future. And using Blockchain is undoubtedly the best way to introduce the benefits of devolution to cloud computing. The need for a decentralized network in modernizing cloud computing architecture has been acknowledged by developers all around the world. The IoT market could gain a lot from the integration of distributed ledger technology and artificial intelligence. The idea of edge computing is comparable. In edge computing, some data is displaced from one or more central nodes to end-users or other edge local cloud, grid, fog computing, distributed data storage, and other more advanced solutions. It is a hybrid of peer-to-peer (P2P), ad hoc networking, local cloud, grid, fog computing, distributed data storage, and other more advanced solutions.

WHAT SOLUTIONS WOULD BLOCKCHAIN OFFER FOR THESE ISSUES?

Because it is built on a distributed ledger system, the blockchain cannot be created without the consent of all stakeholders. Blockchains also have the advantage of being immutable, meaning that once data is placed on one, it is there for good. As a result, you can ensure the security of your assets and data kept in the cloud and know where, when, and how your data was used. Therefore, we can all agree that blockchain addresses some, if not all, of the key issues with cloud computing. Blockchain technology has already shown its value with Bitcoin in the world of financial cryptocurrencies, prompting interest from a wide range of other businesses. Therefore, cloud computing will soon imitate the financial sector's example. Additionally, blockchain technology with its incentive system can be useful. The hosts of different nodes can be compensated with tokens, which serve as the network's currency. It's a successful strategy for encouraging interested people to pool their resources and discourage dishonest individuals. With so many sugar cubes, there are some sour places. The institution's integrity will be questioned if a payment system is not in place to give stakeholders a sufficient incentive structure. It might also prevent the distribution of unfair incentives. Scalability, another attribute of cloud computing, is its capacity to handle enormous data volumes. How will Blockchain be able to gather enormous amounts of data, show scalability, and guarantee the effectiveness and speed of big transactions and data transfers given the existing infrastructure systems in place and their flaws? One of the most important characteristics is that the program is properly computed, which makes it less vulnerable to attacks from cybercriminals. It is the worst kind of failure, leading to a serious breach in the confidentiality of sensitive data that, if tampered with, might be exploited to devastating effect. This necessitates placing a focus on improving its security and accuracy in order to remove discrepancies. If a supplier cannot ensure the protection of these fundamental rights, they inadvertently withdraw their offer to provide the services. It's crucial to develop a trustworthy interface that links smart contracts to actual events because a failure in this system could result in a catastrophic calamity. Erasure control is another issue that needs to be handled if and when the wall falls. regulating the amount of data that is deleted, determining what data should be preserved, and, in some cases, keeping a reliable backup. These abilities also need to be distributed well in order to reduce significant losses.

CONCLUSION

Certainly, cloud computing has caught our attention. No more wasting time and labor by transferring vast amounts of data across different devices. You have Google or Apple accounts for this reason. All of your files will be visible just simply logging in. While there is no doubt that it has made our lives easier, the increased complexity of our lives calls for more original approaches to time management, efficiency, and control. Many of these issues are existential as a result of centralized systems and servers. Blockchain has emerged as the game-changer after extensive research to find the answers. True, there is a to-do list that has to be accomplished before blockchain can be properly integrated with Cloud Computing, despite all the advantages it offers. There is still work to be done.

REFERENCES

Adeniyi, E. A., Ogundokun, R. O., Misra, S., Awotunde, J. B., & Abiodun, K. M. (2022). Enhanced security and privacy issue in multi-tenant environment of green computing using blockchain technology. In *Blockchain Applications in the Smart Era* (pp. 65–83). Springer International Publishing. doi:10.1007/978-3-030-89546-4_4

Ahmed, I., Zhang, Y., Jeon, G., Lin, W., Khosravi, M. R., & Qi, L. (2022). A blockchain-and artificial intelligence-enabled smart IoT framework for sustainable city. *International Journal of Intelligent Systems*, *37*(9), 6493–6507. doi:10.1002/int.22852

Al-Turjman, F., & Salama, R. (2021a). Cyber security in mobile social networks. In Security in IoT social networks (pp. 55-81). Academic Press.

Al-Turjman, F., & Salama, R. (2021b). Security in social networks. In *Security in IoT Social Networks* (pp. 1–27). Academic Press.

Al-Turjman, F., & Salama, R. (2020). An Overview about the Cyberattacks in Grid and Like Systems. *Smart Grid in IoT-Enabled Spaces*, 233-247.

Alromaihi, N., Ismail, Y., & Elmedany, W. (2022, October). Literature Review of Blockchain-based Cloud Computing: Data Security Issues and Challenges. In *2022 International Conference on Data Analytics for Business and Industry (ICDABI)* (pp. 484-492). IEEE. 10.1109/ICDABI56818.2022.10041637

Fateh, A. M., & Salarnejad, A. A. (2022). The Scope of Blockchain Technology: A Meta-Synthesis Study of Applications, Benefits, Challenges, and Related Technologies. *Information & Management*, *8*(1), 245–300.

Habib, G., Sharma, S., Ibrahim, S., Ahmad, I., Qureshi, S., & Ishfaq, M. (2022). Blockchain Technology: Benefits, Challenges, Applications, and Integration of Blockchain Technology with Cloud Computing. *Future Internet, 14*(11), 341. doi:10.3390/fi14110341

Han, R., Yan, Z., Liang, X., & Yang, L. T. (2022). How can incentive mechanisms and blockchain benefit with each other? a survey. *ACM Computing Surveys, 55*(7), 1–38. doi:10.1145/3539604

Kaur, A., Singh, G., Kukreja, V., Sharma, S., Singh, S., & Yoon, B. (2022). Adaptation of IoT with blockchain in Food Supply Chain Management: An analysis-based review in development, benefits and potential applications. *Sensors (Basel), 22*(21), 8174. doi:10.3390/s22218174 PMID:36365871

Khanna, A., Sah, A., Bolshev, V., Burgio, A., Panchenko, V., & Jasiński, M. (2022). Blockchain–Cloud Integration: A Survey. *Sensors (Basel), 22*(14), 5238. doi:10.3390/s22145238 PMID:35890918

Nguyen, T., Vo, K. T., Ta, T. T., Nguven-Hoang, T. A., & Dinh, N. T. (2022, October). The benefits and challenges of applying Blockchain technology into Big Data: A literature review. In *2022 International Conference on Advanced Technologies for Communications (ATC)* (pp. 219-224). IEEE. 10.1109/ATC55345.2022.9943004

Patan, R., Parizi, R. M., Dorodchi, M., Pouriyeh, S., & Rorrer, A. (2023). *Blockchain Education: Current State, Limitations, Career Scope, Challenges, and Future Directions.* arXiv preprint arXiv:2301.07889.

Qamar, R., & Zardari, B. A. (2023). A Study of Blockchain-Based Internet of Things. *Iraqi Journal For Computer Science and Mathematics, 4*(1), 15–23.

Rabbani, M. R., Hassan, M. K., Hudaefi, F. A., & Shaikh, Z. H. (2022). Islamic finance and cryptocurrency: a systematic review. *FinTech in Islamic Financial Institutions: Scope, Challenges, and Implications in Islamic Finance,* 279-306.

Salama, R., & Al-Turjman, F. (2022, August). AI in Blockchain Towards Realizing Cyber Security. In *2022 International Conference on Artificial Intelligence in Everything (AIE)* (pp. 471-475). IEEE. 10.1109/AIE57029.2022.00096

Sathiya, V., Nagalakshmi, K., Jeevamalar, J., Babu, R. A., Karthi, R., Acevedo-Duque, Á., ... Ramabalan, S. (2023). Reshaping healthcare supply chain using chain-of-things technology and key lessons experienced from COVID-19 pandemic. *Socio-Economic Planning Sciences, 85*, 101510. doi:10.1016/j.seps.2023.101510 PMID:36687377

Shah, K., Patel, N., Thakkar, J., & Patel, C. (2022). Exploring applications of blockchain technology for Industry 4.0. *Materials Today: Proceedings*, *62*, 7238–7242. doi:10.1016/j.matpr.2022.03.681

Chapter 5
Analysis and Applications Finding of Wireless Sensors and IoT Devices With Artificial Intelligence/Machine Learning

R. Ramya
https://orcid.org/0000-0002-8071-9343
SRM University, India

ABSTRACT

Wireless sensor devices and IoT (internet of things) devices are two interconnected technologies that have revolutionised the way we interact with our surroundings. These devices are at the forefront of the digital transformation and are instrumental in creating smart and connected environments. This chapter first discussed the basics of wireless sensor devices and IoT devices, with pros and cons. Second, it discussed the major differences between the two; third, the major real-time applications of both types of devices are discussed, and fourth, the discussion about the use of AI and ML algorithms in these devices with detailed explanations of major algorithms are presented. Finally, there is a discussion about the future scope of both wireless sensor devices and IoT devices with AI and ML. Through this chapter, one can learn about different kinds of devices for different applications with various AI and ML algorithms. The future smart world requires many more smart devices for various applications. This chapter can provide a clear understanding of various devices and applications for the future smart world.

DOI: 10.4018/979-8-3693-0786-1.ch005

1. INTRODUCTION ABOUT THE WIRELESS SENSOR DEVICES AND IOT DEVICES

Wireless sensor devices are small electronic devices equipped with sensors that can measure and collect data from their surroundings. These sensors can detect various parameters such as temperature, humidity, light, motion, pressure, and more. They are designed to be wirelessly connected, enabling seamless communication and data transmission (Ghate & Vijayakumar, 2018).

IoT devices, on the other hand, are a broader category that encompasses a wide range of interconnected objects, systems, and devices. These devices are embedded with sensors, software, and network connectivity, allowing them to exchange data and interact with each other and the cloud.

The combination of wireless sensor devices and IoT devices has unlocked tremendous opportunities in various fields, including home automation, industrial automation, agriculture, healthcare, transportation, and smart cities. These devices form the backbone of the ever-expanding network of connected devices that make up the IoT ecosystem.

The wireless connectivity of these devices enables real-time data collection, analysis, and remote control. The data gathered from these devices can be used to gain insights, monitor environments, optimise processes, and make informed decisions. For example, in agriculture, wireless sensor devices can monitor soil moisture levels and weather conditions, enabling farmers to optimise irrigation and resource allocation.

Furthermore, IoT devices can be remotely managed and controlled through centralised platforms or mobile applications. This allows for efficient monitoring, troubleshooting, and automation of various tasks. For instance, in a smart home setup, IoT devices such as smart thermostats, lighting systems, and security cameras can be controlled and monitored from a smartphone.

However, the widespread adoption of wireless sensor devices and IoT devices also raises concerns regarding data security, privacy, and interoperability. As these devices collect and transmit sensitive data, robust security measures must be in place to protect against unauthorised access and data breaches.

Wireless sensor devices and IoT devices have transformed the way we interact with our environment, providing us with valuable data and enabling advanced automation and control. With ongoing advancements in technology and increasing connectivity, these devices will continue to shape the future of various industries, fostering efficiency, sustainability, and an improved quality of life.

2. PROS AND CONS OF WIRELESS SENSOR DEVICES AND IOT DEVICES

Wireless sensor devices and IoT devices both have their own set of advantages and disadvantages. Here are some of the key pros and cons of each:

Wireless Sensor Devices

Pros

Low power consumption: Wireless sensor devices are often designed to operate on low power, making them suitable for long-term monitoring applications.
Cost-effective: Compared to IoT devices, wireless sensor devices are often less expensive to deploy.
Simple to use: Wireless sensor devices are typically easy to install and use as they are designed for a specific sensing task.

Cons

Limited functionality: Wireless sensor devices are often limited in terms of the data they can collect and the processing they can perform.
Limited connectivity: Wireless sensor devices are often designed to communicate with a local hub or receiver and may not be able to connect to the internet or other devices.
Limited data processing: Wireless sensor devices often require a central processing system to analyse the data they collect.

IoT Devices

Pros

Advanced functionality: IoT devices are often more complex than wireless sensor devices and can perform more advanced functions such as running applications and interacting with other devices.
Connectivity: IoT devices are designed to connect to the internet and communicate with other devices, making them ideal for remote monitoring applications.
Data processing: IoT devices can process data locally or in the cloud and often have the ability to learn and adapt based on data analysis.

Cons

Higher power consumption: IoT devices often require more power than wireless sensor devices, which can be a disadvantage in long-term monitoring applications.
More expensive: IoT devices are often more expensive than wireless sensor devices, particularly if they require additional connectivity hardware.
More complex to use: IoT devices can be more difficult to install and use, particularly if they require integration with other systems or devices.

In summary, both wireless sensor devices and IoT devices have their own unique strengths and weaknesses, and the choice between the two will depend on the specific needs and requirements of the application.

3. DIFFERENCE BETWEEN WIRELESS SENSOR DEVICES WITH IOT DEVICES

Wireless sensor devices and IoT devices are both used to collect data, but there are some key differences between them:

Connectivity: IoT devices are designed to connect to the internet and communicate with other devices, while wireless sensor devices typically only communicate with a local receiver or hub.
Functionality: IoT devices are often more complex than wireless sensor devices and can perform more advanced functions, such as running applications and interacting with other devices. Wireless sensor devices, on the other hand, are generally simpler and designed to perform a specific sensing task.
Power consumption: Wireless sensor devices are often designed to operate on low power, as they may be deployed in locations where it is difficult to change batteries. IoT devices, on the other hand, are often designed with more powerful processors and may consume more energy.
Data processing: IoT devices can process data locally or in the cloud, and often have the ability to learn and adapt based on data analysis. Wireless sensor devices, on the other hand, are typically designed to send raw data to a central processing system for analysis.

Overall, while both wireless sensor devices and IoT devices can be used to collect data, they are designed for different purposes and have different capabilities. Wireless sensor devices are typically used in industrial settings to monitor specific parameters, while IoT devices are used in a wider range of applications, from smart homes to autonomous vehicles.

Table 1. The summarization of key differences between wireless sensor devices and IoT devices

Feature	Wireless Sensor Device	IoT Device
Purpose	Collect data from the environment	Collect data from a verity of sources
Size	Small and lightweight	Can be of any size
Power consumption	Limited battery life	Cab be powered by a variety of sources
Security	Vulnerable to security threats	Can be more secure if properly configured

Table 2. The different kind of wireless sensor devices and IoT devices with descriptions

Wireless Sensor Device	IoT Device	Description
Temperature sensor	Smart thermostat	Controls the temperature in a home or office
Pressure sensor	Smart tire pressure monitor	Alerts drivers when tire pressure is low
Proximity sensor	Smart door lock	Allows users to lock and unlock doors remotely
Accelerometer and gyroscope sensor	Smart fitness tracker	Tracks steps, distance, and calories burned
IR sensor	Smart security camera	Detects motion and sends alerts
Optical sensor	Smart light bulb	Turns lights on and off automatically
Gas sensor	Smart gas leak detector	Alerts users to gas leaks
Smoke sensor	Smart smoke detector	Alerts users to smoke and fire

Here is Table 1. It summarizes the key differences between wireless sensor devices and IoT devices.

Table 2 describes the different kind of wireless sensor devices and IoT devices details in specific application's names.

Descriptions About Sensor Devices

A wireless sensor network consists of multiple interconnected sensor nodes that communicate wirelessly to collect and transmit data. Each sensor node is equipped with sensing capabilities (e.g., temperature, humidity, light, motion) and wireless communication capabilities. These nodes are typically small, low-power devices designed for specific sensing tasks.

Here's a description of a typical wireless sensor network setup:

Sensor Nodes: These are the individual devices responsible for sensing the environment and collecting data. They are equipped with sensors to measure physical quantities and transceivers for wireless communication. Sensor nodes are often battery-powered and have limited computational resources.

Wireless Communication: Sensor nodes communicate with each other using wireless protocols such as Zigbee, Bluetooth Low Energy (BLE), or Wi-Fi. They form a self-organizing network, enabling data transmission and routing between nodes.

Gateway/Access Point: In a wireless sensor network, there is typically a gateway or access point that acts as a bridge between the sensor nodes and a wider network (e.g., the internet). The gateway collects data from the sensor nodes and forwards it to a centralized server or a cloud-based system for further processing and analysis.

Centralized Server/Cloud Infrastructure: The data collected from the sensor nodes is sent to a centralized server or cloud infrastructure for storage, analysis, and visualization. These systems handle data processing tasks, apply algorithms for monitoring and decision-making, and provide a user interface for data visualization.

Monitoring and Control: The data collected by the sensor nodes can be used for various applications such as environmental monitoring, industrial automation, smart agriculture, or home automation. The collected data is processed and analyzed to extract meaningful information, which can then be used to control actuators or trigger alerts based on predefined rules (Ramya & Ramamoorthy, 2023).

Table 3 shows the details of different sensor devices, use cases with sample images.

4. APPLICATIONS OF WIRELESS SENSORS AND IOT DEVICES

Wireless sensor devices and IoT devices have a wide range of applications across various industries. Here are some examples of how they are being used:

Wireless Sensor Devices

Environmental monitoring: Wireless sensors can be used to monitor air and water quality, temperature, humidity, and other environmental parameters.

Industrial automation: Wireless sensors can be used to monitor equipment performance, detect faults, and optimize processes in manufacturing plants.

Healthcare: Wireless sensors can be used to monitor patient vital signs, track medication usage, and detect falls.

Table 3. Different sensor devices, use cases and sample images

Sensor Device	Use Case	Image
Temperature sensor	Monitors temperature in a variety of environments, such as homes, businesses, and industrial settings. Can be used to control heating and cooling systems, track food safety, and monitor industrial processes.	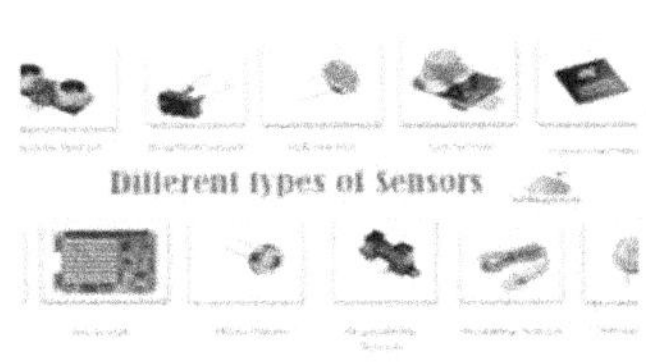
Motion sensor	Detects movement and can be used for security, automation, and other applications. For example, motion sensors can be used to trigger security alarms, turn on lights when someone enters a room, or start a video recording.	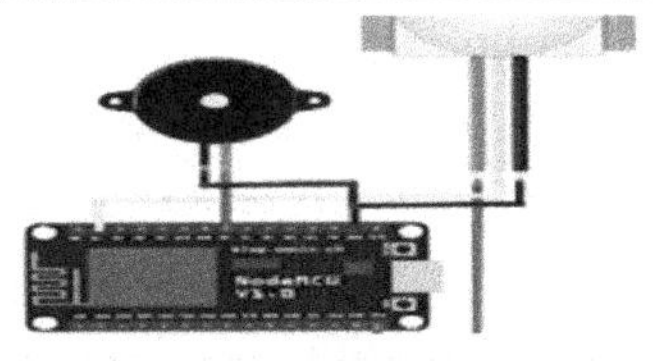
Camera	Captures images and videos. Can be used for security, surveillance, and other applications. For example, cameras can be used to monitor people and property, track inventory, and provide live video feeds.	
Accelerometer	Measures acceleration and can be used to track movement, detect impacts, and measure vibration. For example, accelerometers can be used to track fitness data, monitor vehicle performance, and detect earthquakes.	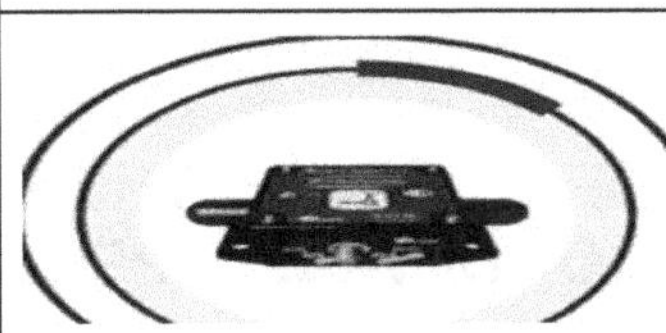
Gyroscope	Measures rotation and can be used to track orientation, detect motion, and measure vibration. For example, gyroscopes can be used to control drones, track gaming performance, and measure the impact of an earthquake.	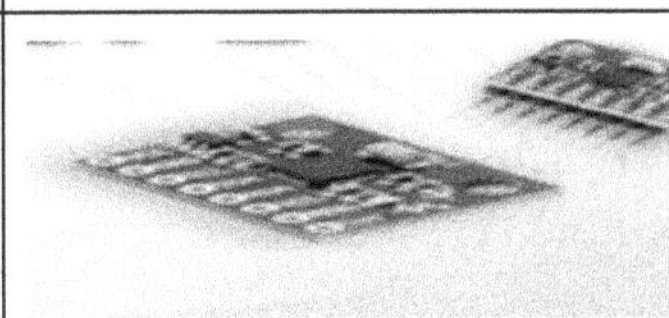
Humidity sensor	Measures humidity and can be used to monitor environmental conditions, such as the humidity in a greenhouse or the humidity in a data center.	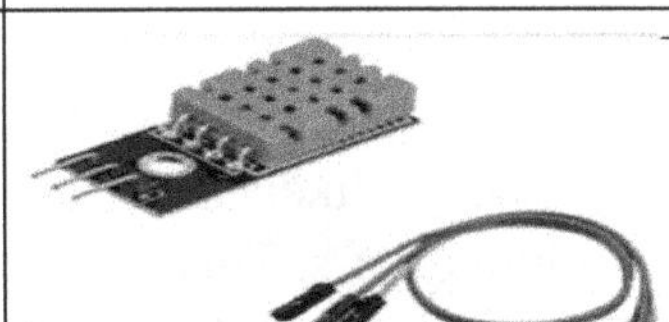
Pressure sensor	Measures pressure and can be used to monitor a variety of applications, such as the pressure in a tire, the pressure in a hydraulic system, or the pressure in the atmosphere.	
Flow sensor	Measures the rate of fluid flow and can be used to monitor a variety of applications, such as the flow of water in a pipe, the flow of oil in an engine, or the flow of air in a ventilation system.	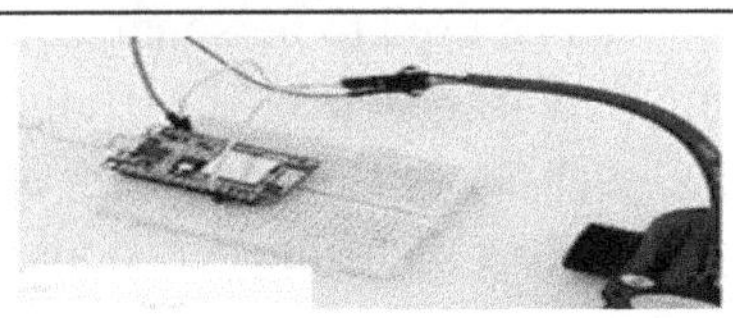

Agriculture: Wireless sensors can be used to monitor soil moisture, temperature, and other environmental factors to optimize crop yield.

IoT Devices

Smart homes: IoT devices can be used to control lighting, heating, and other appliances remotely, and to monitor security and energy usage (Gladence et al., 2020).
Transportation: IoT devices can be used to monitor traffic conditions, optimize route planning, and track vehicles and shipments.
Healthcare: IoT devices can be used to monitor patient health remotely and track medication adherence.
Energy management: IoT devices can be used to monitor energy usage and optimize building heating, ventilation, and air conditioning (HVAC) systems.

Overall, wireless sensors and IoT devices have a wide range of applications, from environmental monitoring to industrial automation to smart homes and beyond. As technology continues to advance, we can expect to see even more innovative uses of these devices in the future (Wang et al., 2019).

5. USE OF AI/ML IN THE WIRELESS SENSORS AND IOT DEVICES APPLICATIONS

AI/ML (Artificial Intelligence/Machine Learning) is increasingly being used in wireless sensors and IoT devices applications to improve their performance and capabilities. Here are some examples of how AI/ML is being used:

Predictive maintenance: AI/ML algorithms can be used to analyze data from wireless sensors and IoT devices to detect patterns and anomalies, predicting when equipment is likely to fail and alerting operators to perform maintenance before a failure occurs. This can help to reduce downtime and maintenance costs.
Energy optimization: AI/ML algorithms can be used to analyze data from IoT devices to optimize energy consumption in buildings, factories, and transportation systems. This can help to reduce energy usage and costs.
Personalization: AI/ML algorithms can be used to analyze data from IoT devices to personalize services and products. For example, data from wearable devices can be used to personalize health and fitness recommendations.
Autonomous systems: AI/ML algorithms can be used to enable autonomous systems, such as self-driving cars and drones, by analyzing data from sensors and making real-time decisions.

Fraud detection: AI/ML algorithms can be used to analyze data from IoT devices to detect fraud in financial transactions or identify security breaches in smart home systems.

Overall, AI/ML is becoming increasingly important in wireless sensors and IoT devices applications, as it can help to improve performance, optimize processes, and provide new capabilities that were previously not possible.

6. WHAT ARE THE AI/ML ALGORITHMS USED FOR WIRELESS SENSORS AND IOT DEVICES APPLICATIONS

There are various AI/ML algorithms used for wireless sensors and IoT devices applications, depending on the specific use case and requirements (Adi et al., 2020). Here are some examples:

Neural networks: Neural networks are a type of machine learning algorithm that are inspired by the structure and function of the human brain. They can be used for a variety of tasks, such as image and speech recognition, natural language processing, and predictive maintenance.

Decision trees: Decision trees are a type of machine learning algorithm that can be used for classification and regression problems. They are often used in predictive maintenance applications, where they can be used to predict equipment failure.

How decision tree algorithm used for wireless sensors and IoT devices applications?

The decision tree algorithm is a machine learning technique that can be applied to various domains, including wireless sensors and IoT (Internet of Things) device applications. Decision trees are particularly useful in these contexts due to their ability to handle complex and dynamic data, make real-time decisions, and provide insights for resource-constrained devices.

Here's a high-level explanation of how decision tree algorithms can be used for wireless sensors and IoT device applications:

1. Data Collection: In wireless sensor networks and IoT devices, data is continuously collected from various sensors or devices. This data can include environmental measurements, device statuses, user interactions, or any other relevant information.
2. Feature Extraction: Before applying the decision tree algorithm, the collected data needs to be preprocessed. This preprocessing step typically involves extracting relevant features from the raw data. For example, if the data includes

temperature, humidity, and light intensity readings, these values can be extracted as features.

3. Labelling: In supervised learning scenarios where the algorithm is trained using labelled data, the collected data needs to be labelled. This means assigning a target or output value to each data point. For instance, in a wireless sensor application, the target could be detecting anomalies or classifying different states of the environment.
4. Training the Decision Tree: Once the data is labelled, it is divided into a training set and a validation set. The training set is used to build the decision tree model. The decision tree algorithm iteratively splits the data based on different features and thresholds to create a tree-like structure that represents a set of rules.
5. Decision Making: After training, the decision tree model can be used for decision making. In wireless sensor and IoT device applications, the decision tree can be employed to make real-time decisions based on the input data received from sensors or devices. For example, the decision tree can determine if the collected data indicates a faulty sensor, abnormal behaviour, or a specific event that requires immediate action.
6. Tree Pruning and Optimisation: Decision trees are prone to overfitting, which means they can become overly complex and perform poorly on new, unseen data. To mitigate this, pruning techniques are applied to simplify the decision tree without sacrificing its predictive power. Additionally, optimisation techniques can be employed to enhance the performance of the decision tree algorithm, such as reducing memory usage or computational complexity.
7. Monitoring and Adaptation: In wireless sensor networks and IoT applications, the environment and data characteristics may change over time. Therefore, the decision tree algorithm needs to be continuously monitored and adapted to ensure its effectiveness and accuracy. This may involve retraining the model periodically using updated data or adapting the decision tree structure dynamically to accommodate changing conditions.

By employing the decision tree algorithm in wireless sensors and IoT device applications, real-time decisions can be made locally without relying on a centralised server, reducing latency and dependency on network connectivity. Decision trees also offer interpretability, as the resulting rules can be easily understood and provide insights into the decision-making process, enabling system optimisation and troubleshooting.

Support vector machines (SVMs): SVMs are a type of machine learning algorithm that can be used for classification and regression problems. They are often used in anomaly detection applications, where they can be used to detect anomalies in sensor data.

Support Vector Machines (SVMs) are indeed used in wireless sensors and IoT device applications. SVMs are a popular machine learning algorithm that can be applied to various classification and regression tasks, including those in wireless sensor networks and the IoT.

Here's how SVMs can be utilised in wireless sensors and IoT device applications:

1. Anomaly Detection: In wireless sensor networks and IoT systems, anomaly detection is crucial for identifying abnormal behaviour, faults, or security breaches. SVMs can be employed to build models that learn the normal patterns of sensor data. The SVM can then classify new data as normal or anomalous based on its proximity to the learned patterns. This helps in detecting deviations or unusual events in real-time, allowing for prompt action.
2. Classification: SVMs can be utilised for classification tasks in wireless sensors and IoT applications. For example, in environmental monitoring, sensors may collect data on different activities or states (e.g., occupancy, temperature levels, air quality). SVMs can learn to classify these states based on the sensor data, enabling tasks such as occupancy detection, object recognition, or identifying environmental conditions.
3. Resource-Constrained Devices: Wireless sensors and IoT devices often have limited computational resources, memory, and power. SVMs can be particularly suitable in such scenarios due to their ability to work well with small training datasets and efficient prediction capabilities. SVMs typically require only a subset of training data called support vectors, which reduces memory requirements and speeds up inference on resource-constrained devices.
4. Non-linear Data: SVMs are known for their ability to handle non-linear data by utilising kernel functions. In wireless sensor networks and IoT applications, the data collected can exhibit complex patterns and relationships. SVMs with appropriate kernel functions can capture these non-linear relationships and make accurate predictions or classifications based on the transformed data.
5. Dimensionality Reduction: In wireless sensor networks and IoT applications, the data collected can have high-dimensional features, leading to the curse of dimensionality. SVMs can be used for dimensionality reduction through feature selection or feature extraction techniques. By selecting the most relevant features or transforming the data to a lower-dimensional space, SVMs can simplify the learning process and improve model performance.
6. Privacy and Security: SVMs can be advantageous in scenarios where data privacy and security are critical. In some wireless sensors and IoT applications, data

may be sensitive and require protection. SVMs can be trained on encrypted or transformed data, enabling privacy-preserving machine learning. This allows for secure model deployment while preserving the confidentiality of sensitive information.

It's important to note that SVMs, like any other machine learning algorithm, should be carefully implemented and tuned according to the specific requirements of wireless sensors and IoT device applications. Factors such as the choice of kernel, hyperparameter tuning, and handling imbalanced data should be considered to achieve optimal performance.

Random forests: Random forests are a type of machine learning algorithm that can be used for classification and regression problems. They are often used in predictive maintenance applications, where they can be used to predict equipment failure.

Random Forest algorithms are commonly used in wireless sensors and IoT device applications. Random Forest is an ensemble learning method that combines multiple decision trees to make predictions or classifications. It has several characteristics that make it suitable for wireless sensors and IoT applications:

1. Robustness: Random Forest algorithms are known for their robustness to noisy and incomplete data. In wireless sensors and IoT systems, the collected data can be prone to noise or missing values due to various environmental factors or sensor limitations. Random Forests can handle such data imperfections and still provide accurate predictions.
2. Classification and Regression: Random Forests can be used for both classification and regression tasks. In wireless sensors and IoT applications, classification may involve tasks such as activity recognition, event detection, or fault diagnosis. Regression tasks can include predicting environmental parameters, energy consumption, or system performance. Random Forests can effectively handle these tasks, providing reliable and accurate results.
3. Feature Importance: Random Forests provide a measure of feature importance, indicating the relevance of different features in the prediction or classification process. In wireless sensor networks and IoT systems, where the number of collected features can be large, understanding feature importance can help in feature selection, dimensionality reduction, or identifying critical factors for decision-making.
4. Non-linearity and Complex Relationships: Random Forests can capture non-linear relationships and complex patterns in the data. In wireless sensor and

IoT applications, where the data can exhibit non-linear behaviour or intricate interdependencies, Random Forests can discover and model these relationships effectively, enabling accurate predictions or classifications.

5. Handling High-Dimensional Data: Random Forests can handle high-dimensional data with ease. In wireless sensor networks and IoT systems, the data collected can have numerous features, making it challenging to analyse and extract meaningful information. Random Forests can handle high-dimensional data efficiently, identifying relevant features and producing reliable results.
6. Outlier Detection: Random Forests can be utilised for outlier detection in wireless sensors and IoT applications. Outliers can indicate faulty sensors, abnormal behaviour, or security threats. By training a Random Forest model on normal data patterns, any deviation from these patterns can be identified as an outlier, allowing for timely action or system maintenance.
7. Real-time Decision Making: Random Forests can provide fast and real-time predictions or classifications. In wireless sensors and IoT systems, where real-time decision-making is often required, Random Forests can process data quickly, allowing for timely responses or actions based on the predictions made by the ensemble of decision trees.

When applying Random Forests to wireless sensors and IoT device applications, it is important to consider hyperparameter tuning, model interpretability, and computational resources. Additionally, domain-specific knowledge and data preprocessing techniques can enhance the performance of Random Forest algorithms in these contexts.

Deep learning: Deep learning is a type of neural network that is used for complex tasks, such as image and speech recognition. It is often used in IoT applications, such as autonomous vehicles and drones.

Deep learning algorithms have found various applications in wireless sensors and IoT devices, enabling advanced data analysis and decision-making. Here are some ways deep learning algorithms are used in these contexts:

1. Sensor Data Analysis: Deep learning algorithms, such as deep neural networks (DNNs), can process and analyse complex sensor data collected by wireless sensors and IoT devices. DNNs can learn hierarchical representations of the data, automatically extracting relevant features and patterns from raw sensor measurements. This enables tasks such as signal processing, event detection, anomaly detection, and fault diagnosis.

2. Image and Video Processing: Wireless sensors and IoT devices with cameras or visual sensors can leverage deep learning algorithms for image and video processing tasks. Convolutional Neural Networks (CNNs) are commonly used for tasks such as object recognition, object tracking, scene understanding, and activity recognition. These algorithms can be applied to applications like surveillance systems, smart home devices, or the visual monitoring of industrial processes.
3. Time Series Analysis: Deep learning algorithms are effective in analysing time series data commonly generated by wireless sensors and IoT devices. Recurrent Neural Networks (RNNs) and their variants, such as long-short-term memory (LSTM) networks, can model temporal dependencies in the data. These algorithms can be used for tasks such as time series prediction, anomaly detection, or event prediction based on historical sensor readings.
4. Sensor Fusion: Wireless sensor networks and IoT systems often consist of multiple sensors collecting data from different modalities. Deep learning algorithms can fuse data from multiple sensors and learn complex relationships between them. This enables the creation of comprehensive models that capture information from different sources, leading to enhanced decision-making capabilities.
5. Energy Optimisation: Deep learning algorithms can optimise energy usage in wireless sensors and IoT devices. By analysing historical sensor data, deep learning models can learn patterns and correlations between sensor measurements and energy consumption. This knowledge can be utilised to develop energy management strategies, including adaptive power allocation, duty cycling, or intelligent energy harvesting, improving the overall energy efficiency of the system.
6. Edge Computing and Real-time Decision-making: Deep learning models can be deployed directly on edge devices, enabling real-time decision-making without relying on cloud connectivity. This is particularly beneficial in wireless sensor and IoT applications where low latency, privacy, or limited network availability are critical. By running deep learning algorithms on edge devices, data can be processed locally, reducing communication delays and preserving privacy.
7. Security and Intrusion Detection: Deep learning algorithms can be employed for security purposes in wireless sensor networks and IoT systems. For instance, recurrent neural networks can analyse network traffic data to detect anomalous patterns indicating cyber-attacks or intrusions. Deep learning models can also be utilised to identify vulnerabilities, detect malware, or protect sensitive data in IoT systems.

It's important to note that deep learning algorithms require large amounts of labelled data and significant computational resources for training. However, advancements in hardware acceleration and optimisation techniques have made it feasible to deploy deep learning models on resource-constrained devices in wireless sensors and IoT applications.

Reinforcement learning: Reinforcement learning is a type of machine learning algorithm that is used for decision-making problems. It is often used in IoT applications, such as autonomous systems, where the algorithm learns through trial and error.

Reinforcement learning (RL) algorithms have the potential to be used in wireless sensors and IoT device applications, although their direct application in these contexts is relatively limited compared to other machine learning techniques. Here are a few ways in which RL algorithms can be employed:

1. Resource Management: RL algorithms can be used for resource allocation and management in wireless sensor networks and IoT systems. For example, in a network of energy-constrained sensors, RL can learn to optimise the allocation of energy resources based on dynamic conditions. The RL agent can adaptively allocate resources to different sensors or tasks, optimising energy usage and prolonging network lifetime.
2. Energy Harvesting and Control: RL algorithms can be used to control energy harvesting and energy consumption in IoT devices. RL agents can learn policies that determine when and how to harvest energy from renewable sources, such as solar panels or kinetic energy, based on environmental conditions and energy demand. This can lead to energy-efficient operation and increased autonomy for IoT devices.
3. Dynamic Spectrum Access: RL algorithms can be employed in wireless communication systems to optimise spectrum access and management. In IoT applications where multiple devices share the same communication medium, RL agents can learn policies for dynamically selecting available channels or adjusting transmission power levels. This can help in reducing interference, improving communication efficiency, and mitigating congestion.
4. Adaptive Sensing and Sampling: RL algorithms can be used to optimise the sensing and sampling strategies in wireless sensor networks and IoT systems. The RL agent can learn to adaptively select sensors, sample rates, or sensing intervals based on task requirements, environmental conditions, or data correlations. This can lead to efficient data collection, reduced energy consumption, and improved data quality.

5. Autonomous Decision-making: RL algorithms can enable autonomous decision-making in IoT systems. For instance, RL agents can learn to make decisions or take actions based on feedback and rewards received from the environment. This can be useful in scenarios where IoT devices need to autonomously adapt their behaviour, such as dynamic environment monitoring, adaptive control systems, or autonomous vehicles.

It's worth noting that RL algorithms typically require extensive exploration and training, which can be challenging for resource-constrained wireless sensors and IoT devices. Additionally, the deployment of RL algorithms in real-world wireless sensor networks and IoT systems may require careful consideration of safety, stability, and adaptability to changing conditions.

While RL has potential in wireless sensors and IoT applications, it is still an active area of research, and its direct application in these contexts is not yet as prevalent as in other domains like robotics or game playing. However, as RL algorithms continue to evolve, we may see more innovative applications of RL in wireless sensor networks and IoT devices in the future.

Ensemble Learning Algorithms: Ensemble learning algorithms combine multiple models to improve overall performance and robustness. Techniques like bagging, boosting, and stacking are used to create ensemble models. Ensemble learning can be beneficial in IoT applications where data may be noisy or exhibit complex patterns.

Ensemble learning algorithms, which combine multiple machine learning models to make predictions or classifications, can be effectively utilised in wireless sensors and IoT device applications. Here are some ways ensemble learning algorithms are used in these contexts:

1. Anomaly Detection: Ensemble learning algorithms, such as Random Forests or Gradient Boosting, can be used for anomaly detection in wireless sensors and IoT systems. By combining the predictions of multiple models, ensemble methods can improve the accuracy and robustness of anomaly detection by identifying abnormal behaviour, faults, or security breaches in real-time.
2. Classification and Regression: Ensemble learning algorithms can enhance classification and regression tasks in wireless sensors and IoT applications. For example, in object recognition or activity recognition, ensemble methods can combine the outputs of multiple models trained on different subsets of the data or using different feature representations. This can lead to improved accuracy and reliability in complex environments.

3. Sensor Fusion: Wireless sensor networks and IoT systems often involve multiple sensors collecting data from different sources. Ensemble learning algorithms can fuse data from multiple sensors, combining the outputs of individual models to make more informed decisions. This enables the creation of comprehensive models that capture information from different modalities, leading to enhanced accuracy and robustness.
4. Model Selection and Combination: Ensemble learning algorithms can help in selecting and combining the best models for specific tasks in wireless sensors and IoT applications. By training multiple models using different algorithms, hyperparameter settings, or feature representations, ensemble methods can select the most accurate models or combine the predictions of diverse models to achieve better overall performance.
5. Real-time Decision-making: Ensemble learning algorithms can enable real-time decision-making in wireless sensors and IoT systems. By combining the outputs of multiple models, ensemble methods can generate predictions or classifications rapidly, without the need for complex computations or reliance on centralised servers. This allows for timely and localised decision-making, reducing latency and dependency on network connectivity.
6. Model Adaptation: Ensemble learning algorithms can facilitate model adaptation in dynamic wireless sensor networks and IoT systems. As the environment or data characteristics change over time, ensemble methods can dynamically update or replace models to maintain optimal performance. This adaptability ensures that the ensemble continues to provide accurate predictions or classifications in evolving scenarios.
7. Model Interpretability: Ensemble learning algorithms can provide insights into the decision-making process and improve model interpretability. By combining the outputs of multiple models, ensemble methods can identify consistent patterns or rules that contribute to the final predictions or classifications. This enables a better understanding of the underlying data and facilitates troubleshooting or optimisation of the wireless sensor or IoT system.

Ensemble learning algorithms, such as Random Forests, Gradient Boosting, or Bagging, can be advantageous in wireless sensors and IoT applications due to their ability to handle complex data, improve accuracy, and provide robust predictions or classifications. However, it's important to carefully select the appropriate ensemble method and consider factors such as model diversity, training data, and computational resources to achieve optimal performance in specific wireless sensor and IoT contexts.

Online Learning Algorithms: Online learning algorithms are designed to learn and adapt in real-time as new data becomes available. These algorithms are suitable for scenarios where data arrives in a continuous stream, such as sensor data in IoT applications. Online learning algorithms enable dynamic model updates and can be used for tasks like real-time anomaly detection and adaptive resource management.

Online learning algorithms have several applications in wireless sensors and IoT devices. Online learning refers to a learning paradigm where models are continuously updated with new data, allowing for real-time adaptation and decision-making. Here are some ways online learning algorithms are used in these contexts:

1. Adaptive Models: Online learning algorithms enable models to adapt and learn from streaming data in real-time. In wireless sensors and IoT applications, where data is continuously generated, online learning allows models to update and incorporate new information without requiring retraining on the entire dataset. This adaptability enables timely responses to changing conditions and ensures that the models remain up-to-date.
2. Sensor Calibration: Online learning algorithms can be used to continuously calibrate sensor measurements in wireless sensor networks and IoT systems. By learning the discrepancies between different sensors or compensating for sensor drift, online learning models can dynamically adjust sensor readings to improve accuracy and reliability. This is particularly useful in applications where precise measurements are essential, such as environmental monitoring or industrial control systems.
3. Anomaly Detection: Online learning algorithms can be applied for real-time anomaly detection in wireless sensors and IoT devices. By continuously updating the model with new data, online learning methods can identify deviations from normal behaviour or detect anomalies as they occur. This enables early detection of faults, security breaches, or abnormal events, allowing for prompt action or system maintenance.
4. Resource Management: Online learning algorithms can optimise resource management in wireless sensor networks and IoT systems. By continuously learning from data, these algorithms can dynamically allocate resources, such as energy, bandwidth, or computation power, based on changing demands or conditions. This leads to efficient resource utilisation, an extended system lifetime, and improved performance.
5. Adaptive Control: Online learning algorithms can be utilised for adaptive control in wireless sensors and IoT applications. By continuously learning and updating control policies based on feedback from the environment, online learning models can adaptively adjust control parameters or actions in response

to changing system states. This allows for dynamic control, optimisation, and adaptability in various IoT applications.

6. Privacy-Preserving Learning: Online learning algorithms can be employed for privacy-preserving learning in wireless sensors and IoT systems. By training models directly on edge devices or local servers, sensitive data can be kept locally, minimising privacy risks associated with transmitting data to a central server. Online learning methods can ensure that models are continually updated while preserving the privacy of sensitive information.
7. Data Reduction and Compression: Online learning algorithms can reduce data dimensionality and compress sensor readings from wireless sensors and IoT devices. By learning compact representations or summarising the incoming data stream, online learning models can reduce the storage and communication overhead associated with transmitting raw data. This is beneficial in resource-constrained scenarios or when bandwidth constraints are a concern.

Online learning algorithms provide the advantage of adaptability, enabling wireless sensors and IoT devices to continually learn from new data and make informed decisions in real-time. However, careful consideration should be given to factors such as model stability, drift detection, and computational efficiency when applying online learning algorithms to wireless sensor networks and IoT systems.

Overall, the choice of AI/ML algorithm will depend on the specific requirements of the application, such as the type of data being collected, the desired outcome, and the available computing resources.

7. WHAT ARE THE FUTURE SCOPE OF WIRELESS SENSORS AND IOT WITH AI/ML

The future of wireless sensors and IoT with AI/ML is vast and promising. Here are some potential future scope and applications:

Autonomous systems: Wireless sensors and IoT devices with AI/ML capabilities will enable the development of more autonomous systems, such as self-driving cars, drones, and robots.

Predictive maintenance: AI/ML algorithms will become more advanced, allowing for more accurate predictions of equipment failure, which will reduce downtime and maintenance costs.

Energy optimization: Wireless sensors and IoT devices with AI/ML capabilities will enable real-time energy optimization in buildings, factories, and transportation systems, leading to significant energy savings.

Personalization: AI/ML algorithms will enable more personalized services and products, such as health and fitness recommendations based on data from wearable devices.

Smart cities: Wireless sensors and IoT devices with AI/ML capabilities will enable the development of more intelligent and efficient cities, with better traffic management, waste management, and energy consumption.

Enhanced cybersecurity: AI/ML algorithms will be used to enhance the cybersecurity of wireless sensors and IoT devices, detecting and responding to security threats in real-time.

Smart Cities, Environmental Monitoring, and Healthcare are key areas where this integration is likely to have a significant impact. AI-powered systems can help in early diagnosis, drug discovery, and healthcare resource management. Wireless sensors and IoT devices can be deployed in agricultural fields to monitor soil moisture, temperature, nutrient levels, and crop health. Industrial automation can improve productivity, minimize downtime, and enhance safety. Transportation can use AI/ML algorithms to optimize traffic flow and predict maintenance needs. Smart homes can use AI/ML algorithms to automate tasks and monitor energy usage.

Overall, the future of wireless sensors and IoT with AI/ML is exciting, as it will enable more efficient, intelligent, and personalized systems across various industries, making our lives easier and more sustainable.

Number of Wireless Sensor Applications for Last 5 Years Comparison

Some common and emerging applications of wireless sensor networks (WSNs) that have gained traction in recent years:

- Smart Cities: WSNs are used for monitoring and managing various aspects of urban environments, including traffic flow, air quality, waste management, parking, and energy consumption.
- Environmental Monitoring: WSNs play a crucial role in monitoring environmental parameters such as air quality, water quality, soil conditions, and biodiversity in both urban and natural settings.
- Industrial Automation: WSNs are deployed in industrial settings to monitor and control manufacturing processes, optimize energy usage, detect equipment failures, and enhance overall efficiency.
- Agriculture: WSNs help in precision agriculture by monitoring soil moisture, temperature, humidity, and other relevant parameters, enabling farmers to optimize irrigation, fertilization, and pest control.

- Healthcare: WSNs are used for remote patient monitoring, tracking vital signs, fall detection for elderly care, and monitoring the condition of patients with chronic diseases.
- Structural Health Monitoring: WSNs are employed for monitoring the structural integrity of buildings, bridges, and other infrastructure to detect potential failures or abnormalities.
- Wildlife Tracking: WSNs equipped with sensors are utilized to track and monitor the behavior, migration patterns, and habitat conditions of various animal species.
- Home Automation: WSNs enable home automation systems for controlling lighting, temperature, security systems, and other home appliances.

These are just a few examples, and the field of wireless sensor networks continues to expand with new applications being explored and developed.

Number of IoT Sensor Applications for Last 5 Years Comparison

Over the last five years, the number of IoT sensor applications has been steadily increasing, driven by advancements in technology, decreasing sensor costs, and the expanding need for data-driven insights across various industries. Here are some notable areas where IoT sensors have been widely deployed:

- Smart Homes: IoT sensors are used for home automation, energy management, security systems, and monitoring environmental conditions such as temperature, humidity, and air quality.
- Industrial IoT (IIoT): In manufacturing and industrial settings, IoT sensors are utilised for asset tracking, predictive maintenance, process optimisation, and real-time monitoring of equipment and production lines.
- Agriculture: IoT sensors enable smart farming practises, including soil monitoring, irrigation management, livestock tracking, and crop health monitoring, to enhance productivity and reduce resource waste.
- Healthcare: IoT sensors find applications in remote patient monitoring, wearable health devices, medication management, and asset tracking within hospitals to improve patient care, safety, and operational efficiency.
- Smart Cities: IoT sensors are deployed for traffic management, parking optimisation, waste management, environmental monitoring, and improving overall urban infrastructure efficiency.

- Retail and Supply Chain: IoT sensors enable inventory management, supply chain optimisation, shelf monitoring, and personalised customer experiences through data-driven insights.
- Energy Management: IoT sensors play a vital role in smart grid systems, energy monitoring, demand response programmes, and optimising energy consumption in buildings and industries.
- Transportation and Logistics: IoT sensors are used for fleet management, vehicle tracking, condition monitoring, predictive maintenance, and optimising routes to enhance efficiency and safety.
- Environmental Monitoring: IoT sensors help monitor air quality, water quality, weather conditions, and natural resource management, aiding in environmental conservation and sustainability efforts.
- Safety and Security: IoT sensors are employed in various domains, such as fire detection, video surveillance, access control systems, and emergency response systems, to enhance safety and security measures.

IoT sensor applications have continued to evolve rapidly with new use cases emerging across industries.

Number of Wireless Sensors Production for Last 5 Years Comparison

The production and adoption of wireless sensors have been on the rise in recent years, driven by advancements in technology, increasing demand for IoT (Internet of Things) applications, and the growing need for real-time data monitoring. Wireless sensors are utilized in various industries and applications, including industrial automation, agriculture, healthcare, smart cities, and more.

According to market research reports, the global wireless sensor market has been experiencing significant growth. For example, a report by Grand View Research projected the global wireless sensor market to reach USD 9.28 billion by 2027, growing at a compound annual growth rate (CAGR) of 8.2% from 2020 to 2027. Another report by MarketsandMarkets estimated the wireless sensor market to reach USD 93.86 billion by 2026, growing at a CAGR of 23.30% from 2021 to 2026.

While these reports provide insights into the overall market growth and potential, they do not provide specific production figures for individual years.

Figure 1. Market value

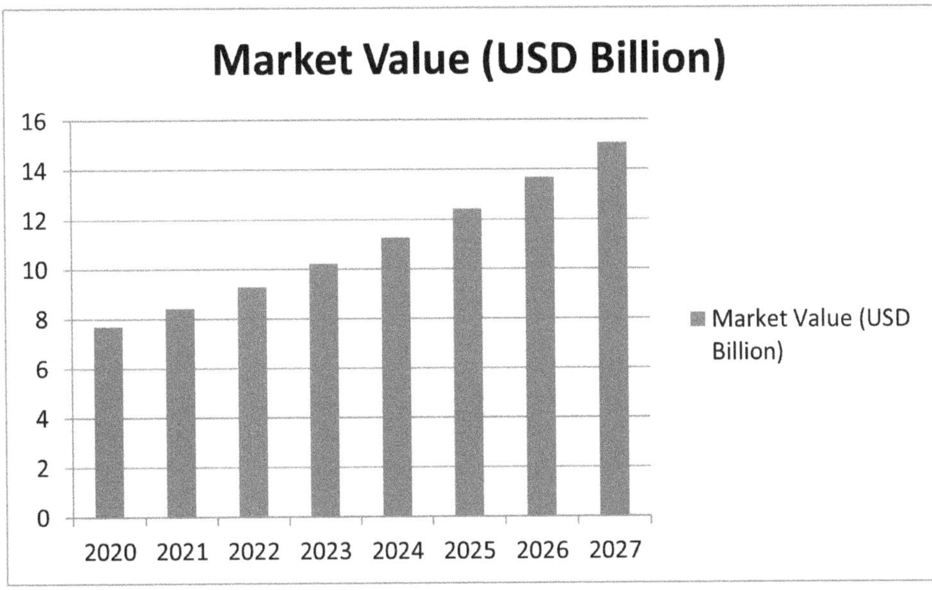

Number of IoT Sensors Production for Last 5 Years Comparison

The production and adoption of IoT sensors have been increasing rapidly in recent years, driven by the growth of the Internet of Things (IoT) and its applications across various industries. IoT sensors are fundamental components of IoT systems, enabling the collection of data from the physical world and facilitating the integration of devices and systems.

According to market research reports, the global IoT sensor market has been experiencing significant growth. For example, a report by Markets and Markets estimated the global IoT sensor market to reach USD 66.98 billion by 2026, growing at a compound annual growth rate (CAGR) of 33.60% from 2021 to 2026. Another report by Grand View Research projected the global IoT sensor market to reach USD 71.62 billion by 2027, growing at a CAGR of 26.2% from 2020 to 2027.

These reports indicate the increasing demand and adoption of IoT sensors across industries. However, they do not provide specific production figures for individual years.

Figure 2. Market value

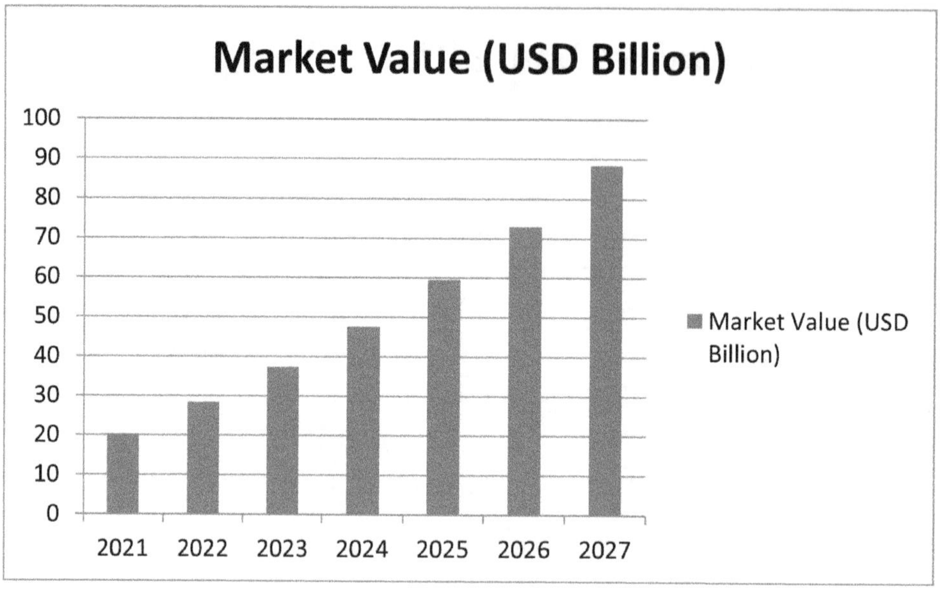

8. CONCLUSION

The integration of wireless sensors, IoT devices, and AI/ML techniques is expected to have a profound impact on future research. It opens up opportunities for data collection, analysis, and automation, enabling more efficient and intelligent systems. This chapter explains the major perspectives of all devices, applications, algorithms, and comparisons for overall understanding. These details will provide a way to find possible future research directions and developments in many application environments.

REFERENCES

Adi, E., Anwar, A., Baig, Z., & Zeadally, S. (2020). Machine learning and data analytics for the IoT. *Neural Computing & Applications*, *32*(20), 16205–16233. doi:10.1007/s00521-020-04874-y

Arjunan, S., & Sujatha, P. (2018). Lifetime maximization of wireless sensor network using fuzzy based unequal clustering and ACO based routing hybrid protocol. *Applied Intelligence*, *48*(8), 2229–2246. doi:10.1007/s10489-017-1077-y

Atzori, L., Iera, A., & Morabito, G. (2010). The internet of things: A survey. *Computer Networks, 54*(15), 2787–2805. doi:10.1016/j.comnet.2010.05.010

Bhatti, G. (2018). Machine learning based localization in large-scale wireless sensor networks. *Sensors (Basel), 18*(12), E4179. doi:10.3390/s18124179 PMID:30487457

Ghate, V., & Vijayakumar, V. (2018). Machine learning for data aggregation in WSN: A survey. *International Journal of Pure and Applied Mathematics, 118*(24), 1–12.

Gladence, L. M., Anu, V. M., Rathna, R., & Brumancia, E. (2020). Recommender system for home automation using IoT and artificial intelligence. *Journal of Ambient Intelligence and Humanized Computing*, (0123456789). Advance online publication. doi:10.1007/s12652-020-01968-2

Ham, H.-S., Kim, H.-H., Kim, M.-S., & Choi, M.-J. (2014). Linear SVM-Based Android Malware Detection for Reliable IoT Services. *Journal of Applied Mathematics, 2014*, 594501. doi:10.1155/2014/594501

Hoomod, H., & Jebur, T. (2018). Applying self-organizing map and modified radial based neural network for clustering and routing optimal path in wireless network. *Journal of Physics: Conference Series, 1003*, 012040. doi:10.1088/1742-6596/1003/1/012040

Kim, W., Park, J., Yoo, J., Kim, H., & Park, C. (2013). Target localization using ensemble support vector regression in wireless sensor networks. *IEEE Transactions on Cybernetics, 43*(4), 1189–1198. doi:10.1109/TSMCB.2012.2226151 PMID:26502429

Kumar, A. (2018). A hybrid fuzzy system based cooperative scalable and secured localization scheme for wireless sensor networks. *International Journal of Wireless & Mobile Networks, 10*(3), 51–68. doi:10.5121/ijwmn.2018.10305

Mahdavinejad, M., Rezvan, M., Barekatain, M., Adibi, P., Barnaghi, P., & Sheth, A. (2017). Machine learning for Internet of Things data analysis: A survey. *Digital Communications and Networks*. Advance online publication. doi:10.1016/j.dcan.2017.10.002

Mottaghinia, Z., & Ghaffari, A. (2018). *Fuzzy logic based distance and energy-aware routing protocol in delay-tolerant mobile sensor networks*. Academic Press.

Phoemphon, S., So-In, C., & Niyato, D. (2018). A hybrid model using fuzzy logic and an extreme learning machine with vector particle swarm optimization for wireless sensor network localization. *Applied Soft Computing, 65*, 101–120. doi:10.1016/j.asoc.2018.01.004

Ramya, R., & Ramamoorthy, S. (2022). Survey on Edge Intelligence in IoT-Based Computing Platform. *Lecture Notes in Networks and Systems*, *356*, 549–561. doi:10.1007/978-981-16-7952-0_52

Ramya, R., & Ramamoorthy, S. (2022). Analysis of machine learning algorithms for efficient cloud and edge computing in the loT. Challenges and Risks Involved in Deploying 6G and NextGen Networks, 72–90.

Ramya, R., & Ramamoorthy, S. (2022). Development of a framework for adaptive productivity management for edge computing based IoT applications. *AIP Conference Proceedings*, *2519*, 030068. doi:10.1063/5.0111710

Ramya, R., & Ramamoorthy, S. (2023). QoS in multimedia application for IoT devices through edge intelligence. *Multimedia Tools and Applications*. Advance online publication. doi:10.1007/s11042-023-15941-6

Ramya, R., & Ramamoorthy, S. (2023). Lightweight Unified Collaborated Relinquish Edge Intelligent Gateway Architecture with Joint Optimization. *IEEE Access : Practical Innovations, Open Solutions*, *11*, 90396–90409. doi:10.1109/ACCESS.2023.3307808

Soni, S., & Shrivastava, M. (2018). Novel learning algorithms for efficient mobile sink data collection using reinforcement learning in wireless sensor network. Wireless Communications and Mobile Computing. doi:10.1155/2018/7560167

Sun, Y., Zhang, X., & Wang, X. (2018). Device-free wireless localization using artificial neural networks in wireless sensor networks. Wireless Communications and Mobile Computing. doi:10.1155/2018/4201367

Wang, J., Gao, Y., Liu, W., Sangaiah, A., & Kim, H. (2019). An improved routing schema with special clustering using PSO algorithm for heterogeneous wireless sensor network. *Sensors (Basel)*, *19*(3), 671. doi:10.3390/s19030671 PMID:30736392

Yadav, A., Kumar, S., & Vijendra, S. (2018). Network life time analysis of WSNs using particle swarm optimization. *Elsevier*, *132*, 805–815.

Chapter 6
Application of Smart Connected-Home Sensors Towards IoT Service

Rupesh Kumar
SRM AP India, India

ABSTRACT

This chapter gives an insight view of the evolving technologies related to the connected home experience using different enabling technologies such as RFID technology as an application of internet of things (IoT) for digitally connected-home facilities. IoT (internet of things), 5G, and the smart wireless network are pushing the real-world interface close to the digital world. Although the gateway of such interface is going to act as a DAC/ADC that is available in electronics designs, such developments are going to bring a better experience of the human-machine establishment by enabling the various technology towards the virtual world with reliable and seamless connectivity. The main purpose of the chapter is to give innovative ways of designing connected-home sensors as an application of IoT services. The IoT is destined to have a profound impact, well beyond the home environment. It will transform virtually all industries, from hospitality and retail to automotive, agriculture, and health sectors.

INTRODUCTION

IoT (Internet-of-Thing) is a network of digitally connected smart devices and sensors. The 5G network and the smart wireless network, IoT, are pushing the real-world interface close to the digital world. Such developments are going to bring a better experience of the human-machine establishment by enabling the various technology

DOI: 10.4018/979-8-3693-0786-1.ch006

towards the Virtual world with reliable and seamless connectivity. MEMS-based smart RFID sensor (presented by Kumar, 2016) is an interesting contributor toward IoT-based applications. In this regard, the development of a Connected-Home (digitally connected) sensors is implemented as an application of IoT. The instantaneous reading through the RFID/NFC-based sensor will easily be monitored by the NFC-enabled mobile. Such digitally connected devices are increasing exponentially with the upgradation of old designs with digital ones. Such as washing machines, vacuum cleaners, door locks, toys and so on which are only limited by the imagination of new designs. This becomes clear as over 3.9 billion were digitally connected in the world in 2016, which was approximately 4.9 million in 2015. In just one year of span, the number has rushed to billions from millions as Gartner reported (a more detailed discussion in this regard is obtained: Gartner/2015). Now, this number is expected to reach 21 billion in 2020. In the line, CISCO has predicted the increase in global IP networks will increase from 16.3 billion in 2015 to 26.3 billion in 2020 (Cisco, 2015). Among this, the Connected Home connections will increase from 2.4 billion in 2015 to 5.8 billion in 2020. Such a huge number of devices will bring more investment in the market. According to the Research Nester (Research Nester, 2016) the global IoT market had reached USD 598.2 billion in 2015 and this is expected to reach USD 724.2 billion by 2023. Such a market is opening several business opportunities for entrepreneurs and start-ups. The IoT is likely to have a huge impact on our daily lives and is going to become an inherent part of the areas like transportation, utility management, healthcare, resource management (electricity, water), and many more. This also brings unprecedented challenges for the researchers and engineers to design more cost-and energy efficient devices/sensors. With so many sensors around us, the security related to personal information needs to be more robust than it would have ever been imagined.

The proposed chapter gives detailed information about some smart sensor designs which are meant to bring in-house human interface with digital appliances much more convenient toward Connected Home. Another important aspect is the incorporation of an Energy Harvesting system in the smart sensor. The researchers are working hard to implement innovative ideas toward this achievement. The practical way to introduce such a system is based on two main concepts; primarily the available energy from the microwave waves is supposed to be trapped for storing, and secondly, based on the voltage coupling concept, the proximity of the sensor to the energy source also makes it possible to harvest. Truly, this makes the IoT service a complete technology for the Connected-Home experience. This chapter also presents the implementation of the RF Energy Harvesting system in alliance with the 5G New Radio guidance.

Figure 1. Attributes of internet of things

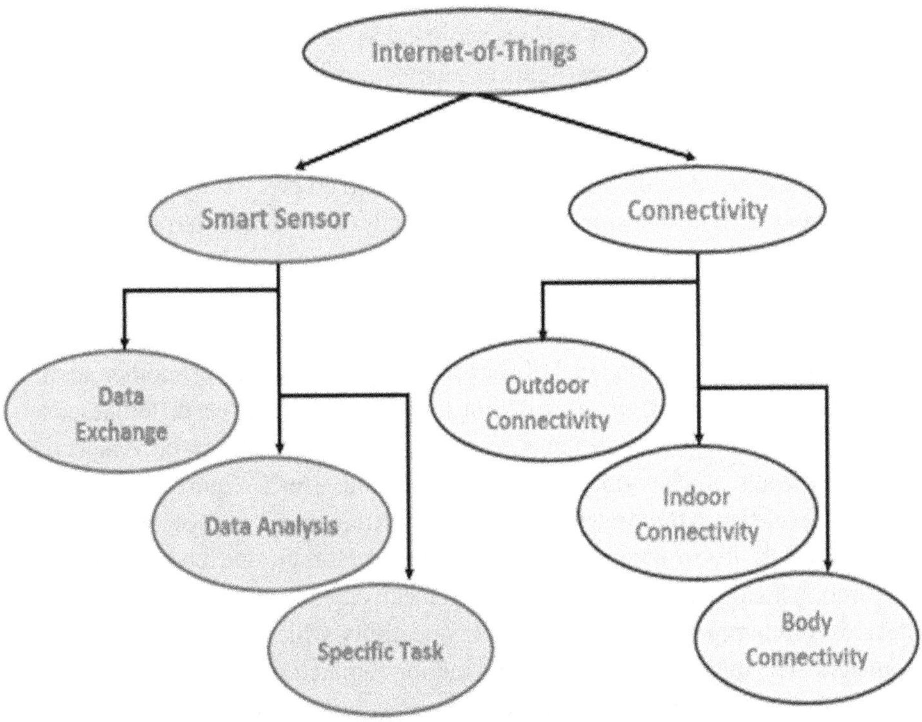

CONNECTED HOME

The meaning of a Connected Home is to remain technologically connected to the home through different smart sensors as a concept of Internet-of-Things (IoT) (Waghmare, Chandgude, Chaure, & Chaudhari, 2017). The two most important attributes of the IoT are 'Smart' and 'Connectivity'. Different attributes of the Internet of Things are depicted in Figure 1.

A sensor or device is assumed to be so (smart) if it has three basic properties: to do data exchange and data analysis and can perform a specific task based on the information obtained by the previous two attributes. Exchanging data with other sensors or devices is the basic requirement to build an IoT. This includes the data generated by the sensors itself as well as the data used to perform the exchange as per the associated protocol of applications or services. Once the data is collected, the analysis is performed for meaningful understanding and further aiding to initiate the task based on the analysis with the minimum assistance from the human. The analysis of data is done at the sensor or device itself if it does not take a long time for computation, for example, temperature sensor. The temperature sensor can determine the surrounding temperature and it can accordingly initiate the task like turning ON or OFF the air conditioner. Such data computation can be easily performed using a microcontroller within the sensor. However, if the analysis involves a massive amount of data, such as managing streetlights or a power grid, then it is not possible to do necessary data analysis at the device level because of heavy computation which is not possible with lower computation-microcontrollers. In such a situation, the data should be transferred to the remotely placed dedicated processing unit to perform rigorous analysis. The decision based on the analysis is transmitted as a command line for the initialization of a specific task. Such computation is generally required for IoT applications implemented in outdoor connectivity such as a smart city. The required computational power depends on the size of its generation by sensors, and hence it also depends on the level of sensor connectivity which is another attribute of IoT. Therefore, connectivity provides more dynamic control over different sensors.

The sensors or devices are designed to adapt to the surroundings as per connectivity at different levels. Such connectivity arises from the user's requirements; this can further be leveled into three parts: outdoor connectivity (a concept of the smart city), indoor connectivity (a concept of the Connected Home), and body connectivity (a wearable smart sensor). The level of connectivities is shown in Figure 2. The outermost circle represents the outdoor connectivity which has an access range in kilometers. The middle circle represents indoor connectivity with access ranges in a few tens of meters, and the inner one represents body connectivity. IoT is meant to connect all the devices or sensors and creates a network of things through cloud access which is marked across the connectivity circles in Figure 2. Similarly, some services such as shopping, and health service are represented across the connectivity level because it covers both body and indoor ranges.

Outdoor Connectivity: The main purpose of this connectivity is to promote or manage the efficient use of public resources such as streetlights, traffic management, monitoring of distribution systems, monitoring of the level of pollution, waste management, and many more. Although the idea of a Smart City is not new, the realization of the idea has started gaining pace with the use of new and more robust connective networks like LoRA, and 5G. This helps in connecting a large number of smart sensors over a wide area, distributed over several kilometers, a network for collecting meaningful information to do data analysis. Normally, a huge amount of data is generated by such a wide network. Real-time analysis of the collected data gives an efficient way of using public resources like streetlight distribution or traffic management. For example, the streetlight will become operational if it senses traffic motion or pedestrian movement. Such an intelligent system helps in utilizing the electricity distribution more effectively as well as it also helps in saving it. Similarly, the

Figure 2. Levels of connectivity

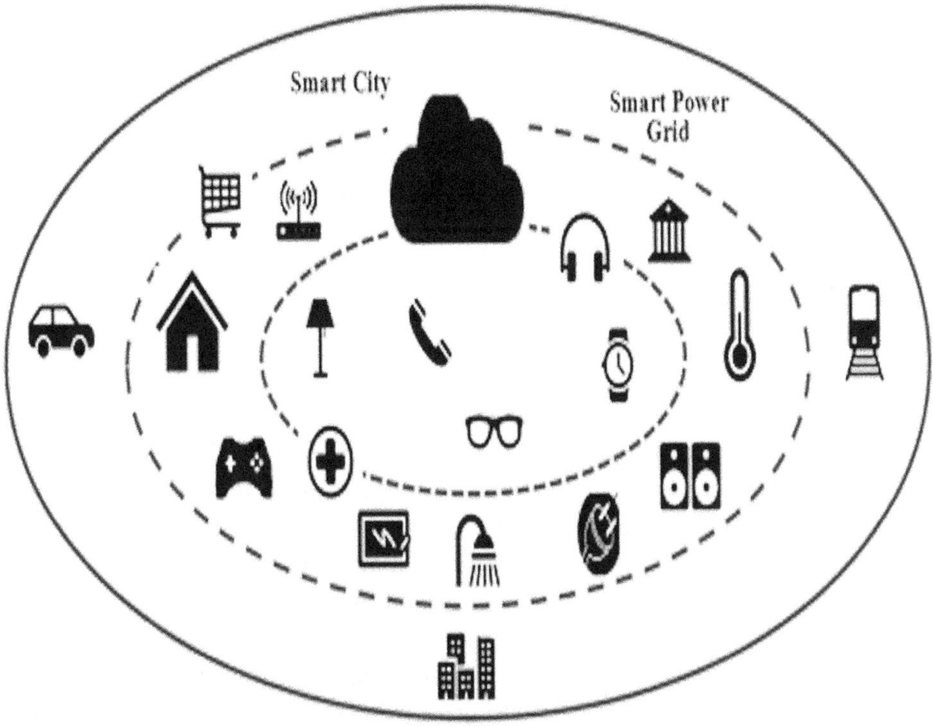

smart power grid can be more efficiently managed by distributing the power as per the requirements in real-time. Economically, the idea of a smart city will lead to the generation of new jobs as it will attract more investment for implementation. This is very much evident from the report of Smart America (Allee, Smart America) which tells that the city governments are going to invest over $41 trillion in the next two decades to upgrade their infrastructure to get benefit from the IoT applications. This intelligent connectivity will surely make life more comfortable in the coming years. The 5G New Radio specifications realized by the 3GPP in Lisbon, in December 2017 are going to be a game changer in the realization of a Smart city with intelligent socio-economic facilities.

Indoor Connectivity: In this level of connectivity, the connected area mainly represents the home level, that is, different home appliances are intelligently connected to form the IoT. Such connectivity will leverage better home management. A smart device or sensor helps in providing data for analysis by different services in real time. Based on the incoming data, the intelligent service will activate the task without the involvement of or minimum intervention by the human.

For example, a smart home can sense the presence and the identity of a person and smart applications helps in learning the normal activities over time. Based on the gathered intelligent information, the smart services will independently make the decision when to turn On or OFF the heater, air conditioner, and lights; activate the security systems; close the door, and windows, or start some intelligent tasks like the initiation of the wireless charging of the smartphone or some other sensors without human intervention. This chapter is mainly dedicated to sensors and applications related to indoor connectivity. Some architecture and design of smart sensors are presented and explained in this chapter within the concept of Indoor Connectivity as Connected Home applications.

Body Connectivity: It represents wearable sensors, that is, IoT at the body level. Compared to the previously mentioned two connectivities, along with the communication interface (such as a smartwatch), it also facilitates the monitoring of health. Such connectivity helps to monitor the well-being of a person by collecting different data related to physical activities (running, walking, sleep), measuring physiological signals, and so. As said, "Prevention is always better than cure". Medicines are the primitive solution to prevent or cure a disease. Although many diseases are cured and prevented with medicine, continuous monitoring of health can prevent many risky diseases. For example, people suffering from diabetes or blood pressure are always advised to take medicine at the time and it is always good to get actively engaged in physical activities as well as following a healthy diet and drinks. However, monitoring so many parameters is not an easy task, especially for old aged people. Here, a wearable sensor or device helps in monitoring the physical activities or it can be used as a reminder alert to follow their medication at the time. A wearable sensor designed to monitor physical activities can help in managing the cholesterol level within the limit and it helps in avoiding many risks like heart attack, stroke. Also, the heartbeat monitoring wearable device can notify the unexpected changes in the heart rate so that the user can take necessary preventive steps or medicines in advance. Apart from health monitoring, the wearable device is also used for communication, for example, a smartwatch helps in checking emails or other notifications. Another important aspect of the wearable smart sensor is to facilitate position information. It helps in tracking pet animals or valuable assets. This also helps in the detection of accidental falls of an old aged person. The fall detection sensor can notify other persons for initiating the care. The mortality rate in old age is high due to undetected accidental falls. In most cases, people did not get medical help at the time because of a lack of such information, and this leads to death which could have been prevented.

The benefits of IoT are limited only by the imagination. Ranging from the outdoors to the human body, the IoT has unlimited applications to make life more comfortable and enjoyable. With 5G, the services are going to bring the virtual world of digital technology close to the real world for a better experience of enjoyment which was not possible a couple of decades back.

APPLICATIONS

According to Statista (Statista, 2018), the number of connected devices (IoT) worldwide will increase from around 23 billion to almost 75.5 billion over 2018-2025. The connected device is going to add more smart features to the home by leveraging intelligent decision-making to the IoT applications with minimum or no intervention from humans. To learn and understand such smart sensor design, this chapter gives a deep knowledge of the fundamental design. The explained sensor diagram can be cost-effectively manufactured. The wireless functionality of sensors is explained by coupling with the RFID technology.

The architecture of an IoT sensor can be broadly classified into three important units: data extraction, data computation, and data exchange as shown in Figure 3.

The extraction of raw data is considered the most important function of any IoT sensor. This is mainly performed by the probe of the sensor which extracts very specific data from the environment. The incoming data (analog) is then converted into digital data with the help of an analog-to-digital converter so that further necessary computation is carried out. At this level of computation, the process is mainly performed related to the reduction of erroneous or ambiguous signals from useful ones. Finally, this data is exchanged with the main processing unit for deep analysis of the collected data to show that necessary action can be taken. The analysis of data can also be done at the computational level itself if it does not require a strong computation and the final analysis can be exchanged with the remote system for further use.

Based on the architecture presented in Figure 3, some examples of sensors (smart water meter, humidity sensor, temperature sensor) meant to use within the connected-home concept are explained in this chapter. Implementation of such a smart meter helps in many ways; a completely autonomous system can be built to perform the real-time reading collection to the bill payments with minimum intervention by the user. This will boost the fast processing, and hence in managing the public distribution system. The two novel low-cost designs presented in this chapter are Water-Meter Sensor and Humidity-Temperature Sensor.

Figure 3. Block diagram of an IoT sensor

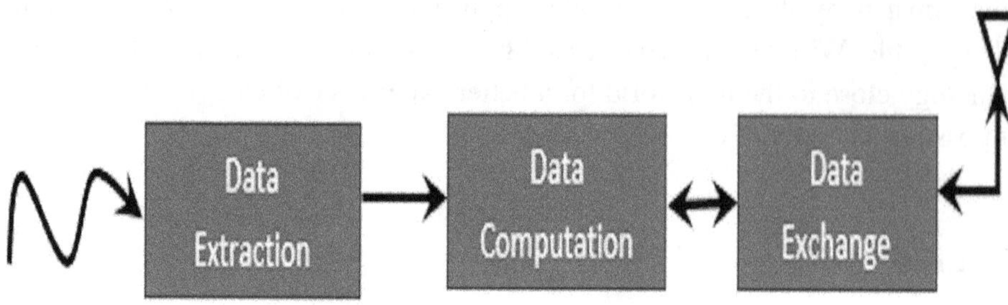

The principle of water flow measurement is based on the difference in time-of-flight (TOF) of an ultrasonic pulse that travels downstream and upstream. The measurement scheme is shown in Figure 4.

The speed of flow is measured as follows,

Where, t_{up}, down are signal travel times TOF_{up} and TOF_{down}

Δt is the time difference in time between tup and t_{down}

L is the ultrasonic signal path length parallel to the water flow

c_s is the speed of ultrasound signal in water

v is the speed of water flow

The speed of water flow (v) is an instantaneous value of the continuous flow at the time of measurement. Therefore, integrating the speed of flow over any certain period gives the total volume of the water flow. The TDC-GP30 is a system-on-chip approach based on the ultrasonic flow converter from ACAM which can be easily used for this purpose.

The block diagram of the water-meter sensor is presented in Figure 5.

The measured data of water flow provided by the TDC-G30 is transferred to a next-generation RFID tag via a microcontroller. Here, the microcontroller acts as master whereas TDC-G30 and RFID tag are slaves. The information written to the tag uses I^2C specification whereas the communication between TDC-G30 and microcontroller is done using SPI specification. The SL3S4011_4021 is a next-generation RFID tag at the UHF band which can be easily used in the above-presented block diagram. The reading of the meter is transferred to a remotely installed RFID reader for further deep analysis over a longer time, that is, for understating and managing the water demands in a house.

The above water-meter design can also be installed using the NFC (Near-Field Communication) technology by replacing the RFID tag and antenna, or as a hybrid system by integrating the NFC chip. For such modification, the NFC chip (M24LR04E-R from ST Microelectronics) can be utilized with the terminal end as

Figure 4. The measurement principle of water flow based on the ultrasonic pulse

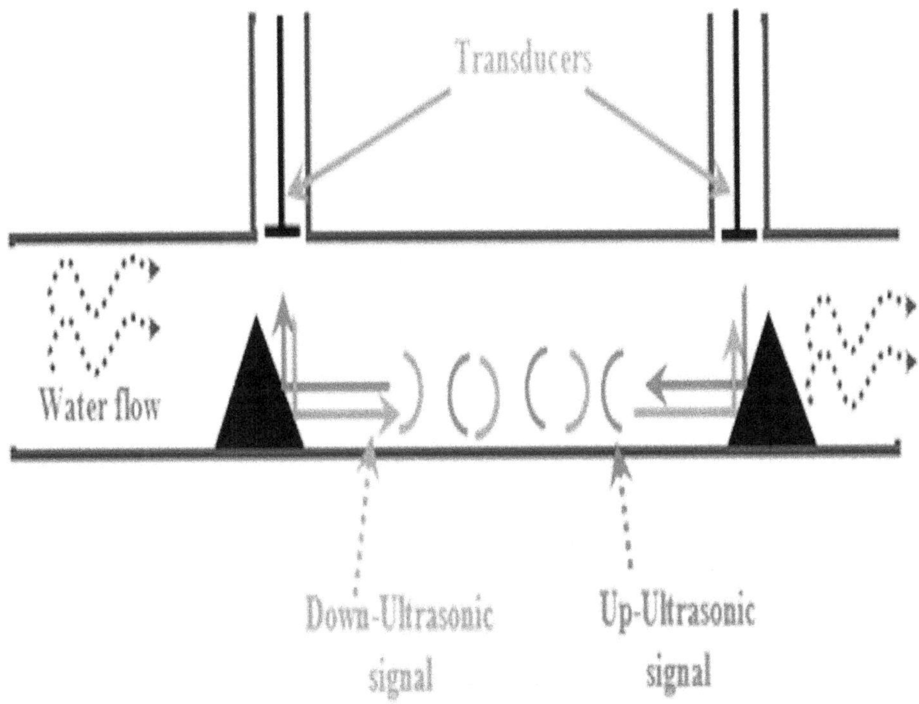

a coil antenna for inductive coupling for NFC communication. This will facilitate the NFC-enabled smartphone to do reading from the water meter. The reason for incorporating such modification is that the NFC technology can be utilized for enabling the ePayment facilities through a smartphone. The instantaneous reading through the NFC-coupled water sensor will easily be monitored by the NFC-enabled smartphone. Further, this specific information along with the user IDs and their corresponding reading will be used to perform the online bill payment over the cellular network. Such a system will give more confidence to the user about the billing and simultaneously it will fasten the processing and the management of a Billing system. With the existing technology, the problem of signal coverage remains a challenge from the implementation point of view. However, such a problem can be avoided with the Connected-Home sensors. Moreover, the MEMS-based sensor coupled with the NFC-RFID technology also enables a cost-effective solution for the monitoring of different electrical home appliances such as electric-meter. Therefore, smart Electric/Water meters can be implemented as an application of IoT towards the development of a Connected-Home (digitally connected). In this

Figure 5. Block diagram of water-meter for IoT application

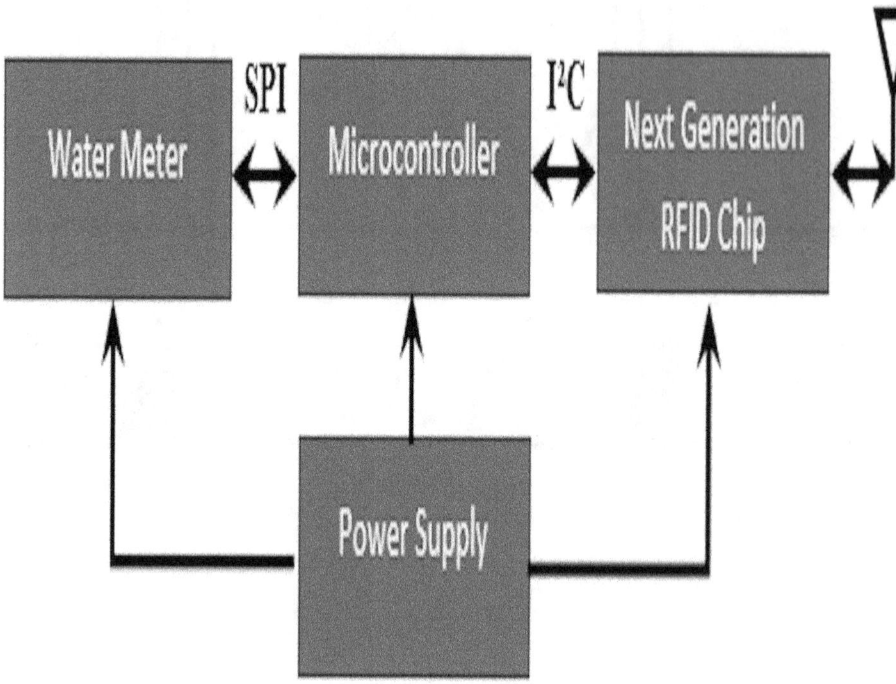

chapter, another example of a smart sensor for humidity and temperature control is described for the reader as an IoT application.

Within the house, humidity and temperature levels should be maintained for health comfort. High humidity is quite uncomfortable if a person is engaged in physical activity, such as exercise. It also increases sweating which is a very common experience. Asthma causes more discomfort in high humidity because the air becomes heavier with moisture, and it becomes more difficult to breathe which results in a high chance of experiencing an asthma attack. On the other hand, low humidity results in dry skin which causes skin itchy and flaky. The experience of dry and rough lips is common because of low humidity. It also makes allergy symptoms worse. However, the use of dehumidifiers and air conditioners can be regulated to maintain the level of comfort. To do so, a low-cost smart sensor design is presented in this chapter. The block diagram of the smart sensor for temperature and humidity measurement inside a house is presented in Figure 6.

The temperature and humidity sensor IC (SHTW2) from Sensirion is a digital humidity and temperature sensor in a flip-chip package. Its overall dimension is 1.3mmx0.7mmx0.5mm mm which is suitable for applications with the tightest space

Figure 6. Block diagram of temperature and humidity sensors based on RFID technology

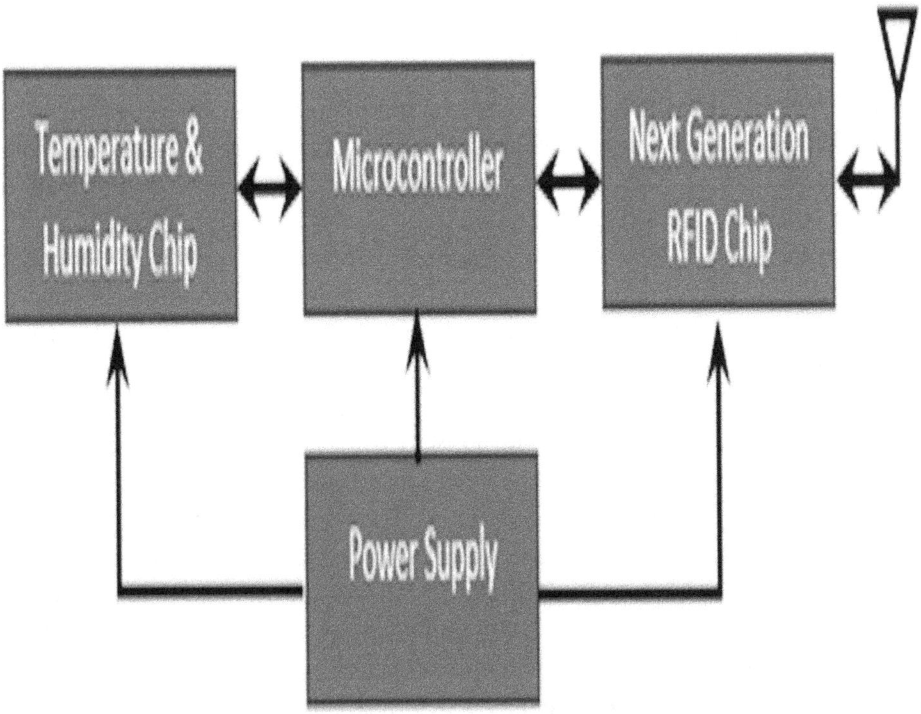

constraints. Its per unit cost is a few US dollars. The SHTW2 uses 16-bit values to measure temperature and humidity. The used conversion formulas are,

Where RH is the relative humidity in percentage, and T is the temperature in degrees Celsius.

S_{RH} and S_T denote the raw sensor output for humidity and temperature, respectively.

The block diagram in Figure 6 uses the SHTW2 sensor for measuring humidity and temperature levels and then the information is wirelessly transferred to the remotely installed UHF RFID reader via RFID tag. The UHF antenna is connected to the tag for the wireless exchange of data.

The SHTW2 acts as a slave for the microcontroller (master) and the commands are exchanged according to the I^2C. The measurement command initiates with a START condition followed by an I^2C header with the 7-bit I^2C device address and a write bit (write W: '0'). The SHTW2 acknowledges the proper reception of the byte by ACK bit (pulling the SDA pin low) after the falling edge of the 8th SCL clock. After this, it is ready to receive the measurement commands from the master (microcontroller). Each time, the SHTW2 acknowledges the proper reception of the byte with the ACK bit. With the acknowledgment of measurement commands, the SHTW2 starts measuring the humidity and the temperature. After the measurement is completed by the SHTW2, the master sends a START condition followed by an I^2C read header. The SHTW2 starts sending the measured data after the acknowledgment of the read command; the two bytes of data are followed by one byte of checksum and each byte is acknowledged by the microcontroller with an ACK condition for the SHTW2 to continue sending data. In the absence of an acknowledgment bit from the microcontroller, the SHTW2 stops sending the measured data. More information about the communication steps can be accessed at Sensiron official website.

The temperature and humidity information can be transferred to a display or a remote base station wirelessly with the use of communication networks such as Bluetooth, WiFi, or RFID. As an example, in this chapter, the long-range RFID technology has been considered to explain the wireless exchange of data between the sensor design and a remotely installed any standard RFID Reader.

The measured data collected from the SHTW2 can be wirelessly transmitted to any standard RFID Reader through the Next Generation RFID Tag such as SL3S4011_4021 from NXP semiconductor. The SL3S4011_4021 also follows the I^2C communication protocol, hence the data is exchanged between the master (microcontroller) and the slave (SL3S4011_4012) likewise mentioned before, between SHTW2 and microcontroller. Once the measured results are transferred to the RFID tag's memory such as EPC (Electronic Product Code) by commanding the WRITE process by the microcontroller, the data will be available for the read by the RFID Reader. The RFID reader transmits the commands by modulating a UHF RF signal. This received by the RFID tag has both information and operating energy, that is, a tag that is passive in nature uses all operating energy from the reader RF waveform. However, the energy (between the microcontroller and the SHTW2) comes from the battery as shown in Figure 3.

Further, based on the received information from SHTW2, the necessary task can be initiated such as the opening of a window or its shutter, and turning ON/OFF air-conditioning. Many uses cases can be used associated with the regulation of humidity and temperature inside a house. Therefore, based on the presented block diagram of the sensor design (Figure 6), many different sensors can be designed for IoT applications. However, the most critical issue that remains with the presented design is the power supply. With time the battery's power gets exhausted and hence it requires a new one. Such power issue is faced by all smart sensors. To solve this, it requires an energy harvesting system that must be able to refill the battery power. RF Energy Harvesting (RFEH) is an interesting way to fulfill the continuous power requirement. In this chapter, the RFEH scheme at mmWave spectrum is presented as prospective applications for the 5G technology.

RF ENERGY HARVESTING

Charging always requires for keeping any electronic devices functional. Traditionally, electronic devices use a charging cable to power up the exhausting battery which is generally performed by the user itself over time as per the level of available power in the electronic devices, for example charging a mobile phone. Today, the charging is mainly performed by connecting the device to a power source through connecting cable a very efficient process. Such a process sometimes becomes a tedious task if the device is not portable like mobile, for example, the sensor described in this chapter for the Connect Home applications. Therefore, if the cable-based charging method is replaced by a wireless method then it will be easy to keep such smart sensors functional without requiring any extra effort from the users. The first development that came in this way was the charging process based on inductive coupling. This involves the transfer of energy from one inductive coil to the receiver's coil. However, this lacks power transfer efficiency, and this also requires very close proximity to perform the charging task. Charging based on the coil induction becomes non-efficient with the separation of some centimeters. Nevertheless, such a method is quite far from wireless charging. To do wirelessly, the available RF energy is used to trap for the harvesting (A. Al-Khayari, H. Al-Khayari, Al-Nabhani, Bait-Suwailam, & Nadir, 2013; Al-Lawati, Al-Busaidi, & Nadir, 2012). Nevertheless, RF Energy Harvesting (RFEH) has some issues which need proper attention from researchers as well as engineers. Some of the issues are mentioned in this chapter as,

i. Availability of Low RF power
ii. Interference with the other waves
iii. Locating and then charging the needed devices

The majority of IoT devices are known to have very low-power energy consumption. For example, an RFID tag mainly requires 10 μW for operation. Although the energy requirements are low, it does require charging with time or the replacement of a battery with a new one. Wireless charging is meant to perform the charging at a distant device. There are some key challenges which are associated with Wireless Charging. Some key challenges are

Key points for designing the RFEH system are as,

1) Low availability of RF power which leads to a longer duration of charging period, and which can be also termed as less efficient in harvesting RF energy
2) Limited availability of spectrum to perform RFEH task
3) RFEH should not introduce any unintentional or unwanted channel interference to the existing communication standards.

As it is well known, the best thing about a Wireless Communication device is that it brings freedom from a connected cable. Likewise, the RF Energy Harvesting system gives freedom from the charging plug-and-cable.

Above mentioned constraints can be properly minimized with the application of the beamforming concept, especially at the mmWave spectrum under the guidance of 5G New Radio guidance. Globally available bands for the 5G implementation are presented in Figure 7.

Between 24 GHz and 30 GHz, the 5G guidance covers all global telecom markets. Therefore, it is interesting to have a harvesting system over 24-30 GHz bands so that it has the advantage of addressing a large number of devices based on 5G guidance with RF energy harvesting. Additionally, the beamforming concept, based on the frontend array design, which is designed for wireless communication at mmWave (5G) can also be explored for the wireless charging task. The beamforming will also provide sufficiently large available RF energy due to the array gain and hence the charging period will be less. This will help in improving the charging rate.

In this chapter, the way to design a wireless RF harvesting system based on the steerable beamforming concept (Sun, Hirata, Ohira, & Karmakar, 2004; Loh & Cheadle, 2017) is discussed with an example of array design.

The beam steering concept should be efficiently used to search for the intended device which is supposed to need electrical power (charging). Once such a device is located and identified, it can directly use to radiate RF signals toward the located (intended) device by providing a directed radiation beam (RF power) based on the beamforming concept. This will bring two advantages, first, it can radiate a high-power RF energy that comes from the array gain, and second, it will avoid unwanted channel interference with other devices. Additionally, at the mmWave spectrum wireless communication is mainly achieved within the Line-of-Sight (LOS) range between two communicating devices compared to the sub-6 GHz frequency spectrum where reflection and diffraction are dominating factors for such communication. At mmWave, the high free path propagation loss occurs compared to sub-6 GHz. Therefore, if the RF beam is oriented to a specific direction for charging purposes, then it is highly unlikely that it will introduce channel interference to the other communicating devices. Hence, the RFEH system based on the steerable beamforming concept at mmWave spectrum has the potential to become a low-cost design. A low-cost RFEH design concept is presented in this chapter which will be easy to implement. The RFEH system can be sub-classified into two main subsystems, a base station, and a harvesting system. The base station directs the RF energy toward an intended receiver, a harvesting system, which stores the incoming RF energy into a reserve such as a battery of the device. The schematic diagram of the RF energy harvesting is presented in Figure 8.

The block diagram of the RF Harvesting System is presented in Figure 9.

Figure 7. Global 5G bands above 24 GHz

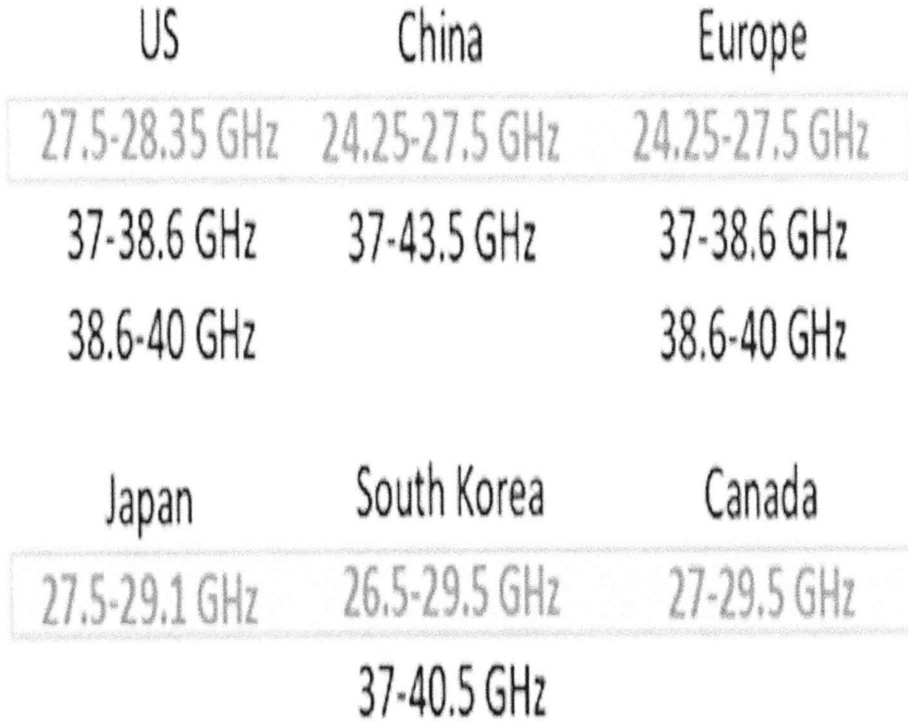

The RF signal received by the antenna is connected to the matching circuit for maximum transfer of energy. The harvesting circuit is made up of non-linear devices such as a diode, because of this non-linearity the circuit also shows non-linear performance. Such non-linear behavior causes variations in the impedance of the energy harvesting circuit with the amount of power received by the receiver's antenna. The maximum power transfer, from the antenna to the harvesting circuit, is achieved only if the circuit 's impedance matches the RF input port of the receiver antenna. Therefore, it is worth having a matching network for the maximum transfer of available RF energy to the storing device such as a battery. More details related to the matching network are explained by the authors (Felini, Merenda, & Corte, 2014).

The RF signal received by the antenna is an AC signal which needs to be converted into a DC signal for storing purposes by a storing device. To convert the Ac signal to a DC signal, a rectifier circuit should be used which consists of simple diodes (Khansalee & Nuanyai, 2015). Once the DC signal is obtained, it is stored in a storage device such as a capacitor.

As the available RF power at the receiver's antenna is mainly affected by the free space path loss in Line-of-Sight (LOS) at the mmWave spectrum. A comparison

Figure 8. Schematic diagram of RF energy harvesting system (RF energy transmitted towards an intended receiver for harvesting)

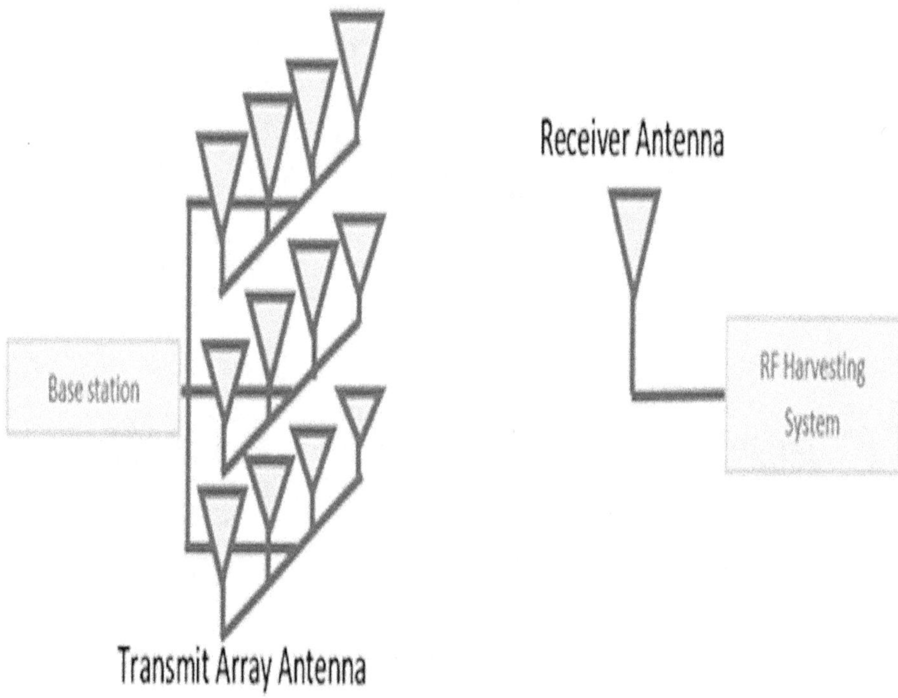

between different available RF energy in terms of antenna gain over distance is depicted in Figure 10.

If the base station has an array gain of 21 dB or more then it will provide more RF energy to the receiver antenna as well as the steerable beam will provide a true automation mechanism for harvesting to an intended device at any spatial distance by searching and identifying. For example, a smart RFID sensor could be charged up by locating and identifying its position through the steerable beam. Therefore, the working concept of the base station is subdivided and explained in two parts 'Search' and 'Charge'. The Search process is responsible for locating the intended device which is achieved by steering the beam through wireless acknowledgment. Once the intended device is found, the base station initiates the charging process by directly radiating the RF energy toward the device acknowledged during the search process. The beam formation provides the extra gain which is directly proportional to the average power available to each element in the array. Therefore, the power attained by a low average powered array can also be achieved with a fewer number of the array by using a high average powered array. The high average power of each

Figure 9. Block diagram of harvesting system

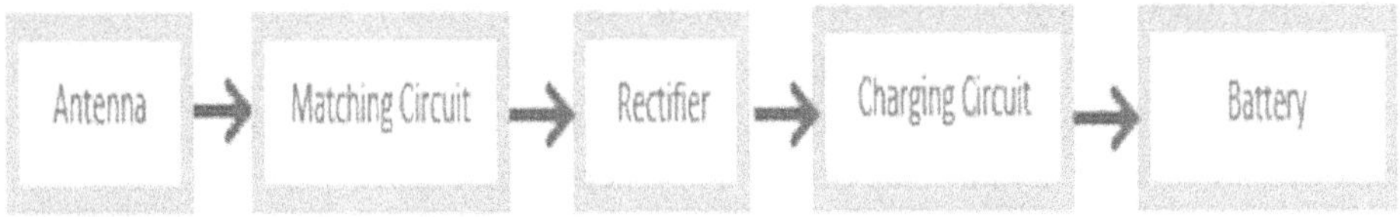

Figure 10. Available RF energy for harvesting with receiver antenna gain

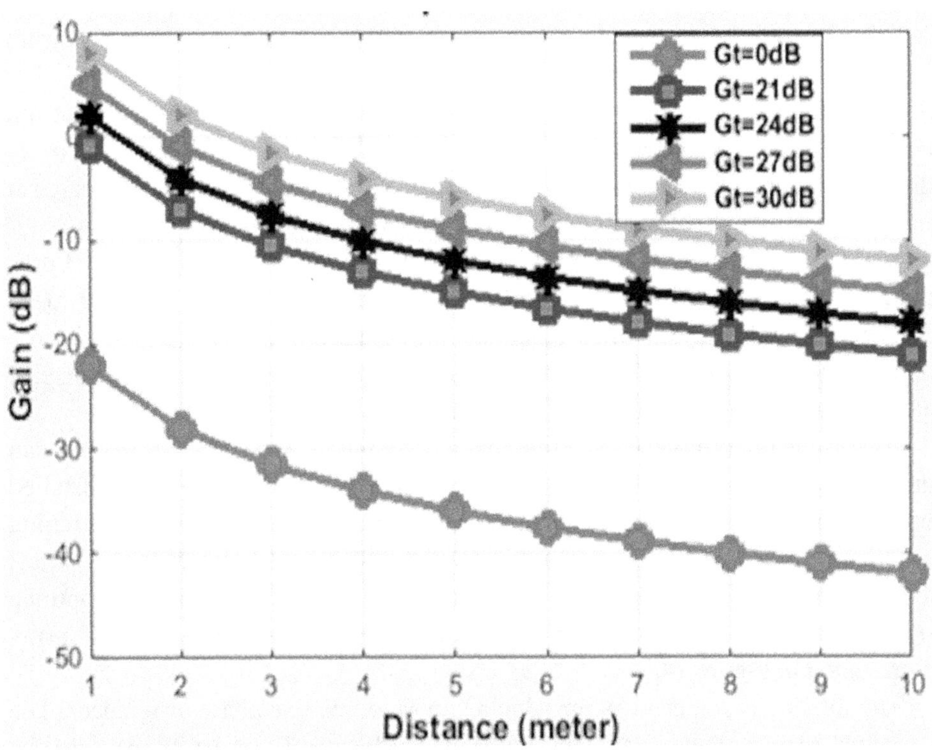

element can be achieved by the high-performance amplifiers and hence this will reduce the overall dimension of the front-end array. However, the use of amplifiers will add up more cost to the final design, therefore, it is economically not a viable solution. It is worth remembering that the base station will be installed in a given area and therefore even if the front-end array antenna size is under 20cm x 20cm (LxW), it will be easily installed. Therefore, the larger array antenna size provides more available RF energy for harvesting than compared to the smaller array.

The application of front-end array antenna gain is visible in Figure 10 which indirectly shows the influence of array size, that is, a larger array provides more available RF energy for harvesting purposes. It is worth mentioning that the gain of the antenna varies with the designs, as an example a stack array antenna design at 28 GHz is presented in this chapter. Stack Antenna Design (L. Wang, J. Wang, & Shi, 2018; Hu, Hao, & Hong, 2017). The simulation is performed in CST and an array of 5x5 stake array antenna design is presented in Figure 11.

The overall dimension of the 5x5 array is 25mm x 25mm x2 mm (LxWxH). The array consists of 25 elementary units of a stack antenna as shown on the right side of Figure 11 and a complete array on the left side. The elementary unit has four layers of substrates. The ground layer is used to excite the upper layers for radiation purposes. The matching bandwidth over the desired mmWave spectrum is presented in Figure 12.

The matching bandwidth is less than -10 dB over the desired mmWave spectrum and it is -30 dB at 26.5 GHz; this represents a bandwidth of 7.8 GHz which covers almost the entire 24 GHz bands for 5G services. The array gain of the 5x5 design at three different frequencies, 28 GHz, 25 GHz, and 31 GHz, are presented in Figure 13.

The 5x5 array gains at 28 GHz, 25 GHz, and 31 GHz are 19.5 dBi, 18.3 dBi, and 20.1 dBi, respectively. The sidelobe level is less than -15.4 dB over the 25-31 GHz frequency band. The array again can attain more gain with the use of more elementary units such as 9x9, and 17x17 arrays. The simulated gain and size of such arrays are tabulated in Table TRFEH1.

As it is clear from the above table that a 65x65 array of size 39cm x 39cm can attain a gain of around 42 dB. Such size of array design, 65x65, can be easily installed inside a room that will do the 'Search' and 'Charge' operations. The searching is performed by steering the beam and once the location of the intended device, supposed to require energy harvesting, is confirmed, the base station will continue to radiate the RF signal towards it. The steerable beam formed by the 17x17 array is presented in Figure 14.

The steering is achieved by introducing constant relative phase differences. The presented steerable beam is obtained with the 60° (degree), 0°, and -60°, respectively. The architecture and beam steering process are explained by the authors (D'Amato, Avitabile, Coviello, & Talarico, 2018; Valdes-Gracia et al., 2018). Such a steerable beam also brings an extra advantage related to the occurrence of any RF channel interference, because the RFEH system uses the directive beamform for energy transfer, therefore, it has low chances of causing channel interference as it occupies narrow beamwidth in space. The beamwidth is proportional to the number of antenna elements as a larger array size gives a narrower beamwidth compared to the fewer size. This reduces the chance of channel interference which is difficult if the charging beam is not directive. It is worth mentioning that this is a critical issue with the RF EH system as undirected are more likely to introduce channel interference.

Figure 11. 5x5 stack array antenna over 24-32 GHz designed from 25 elementary units of stack antenna

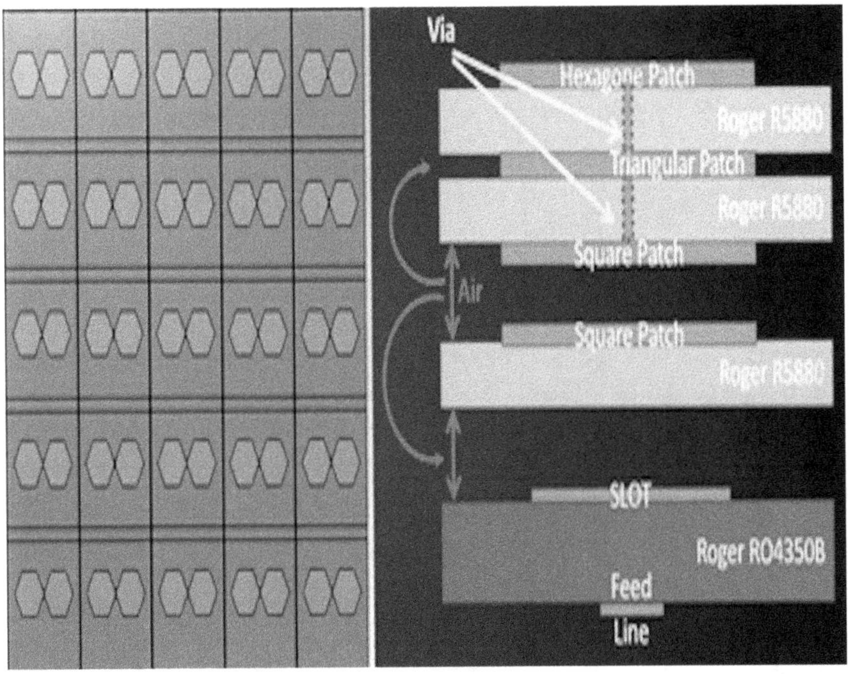

Figure 12. The matching impedance of an elementary unite of stack antenna, 24.1-32 GHz

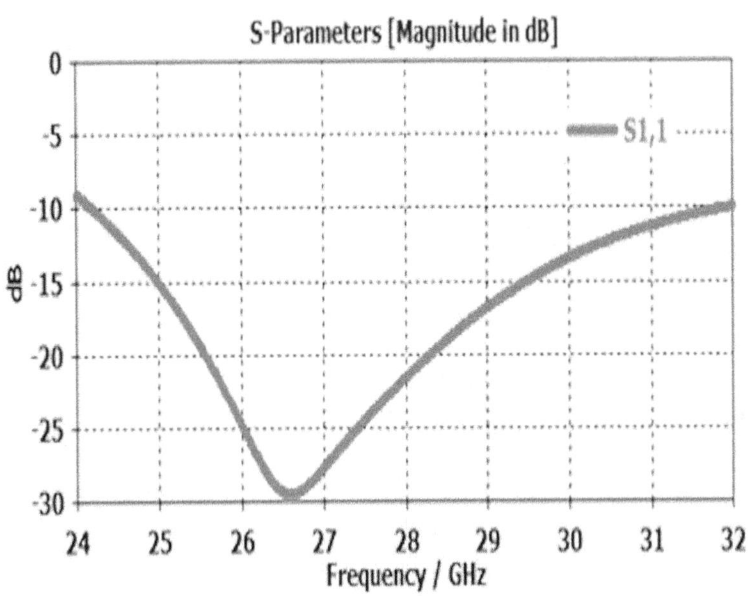

Figure 13. Array gains over at 28 GHz, 25 GHz, and 31 GHz, respectively

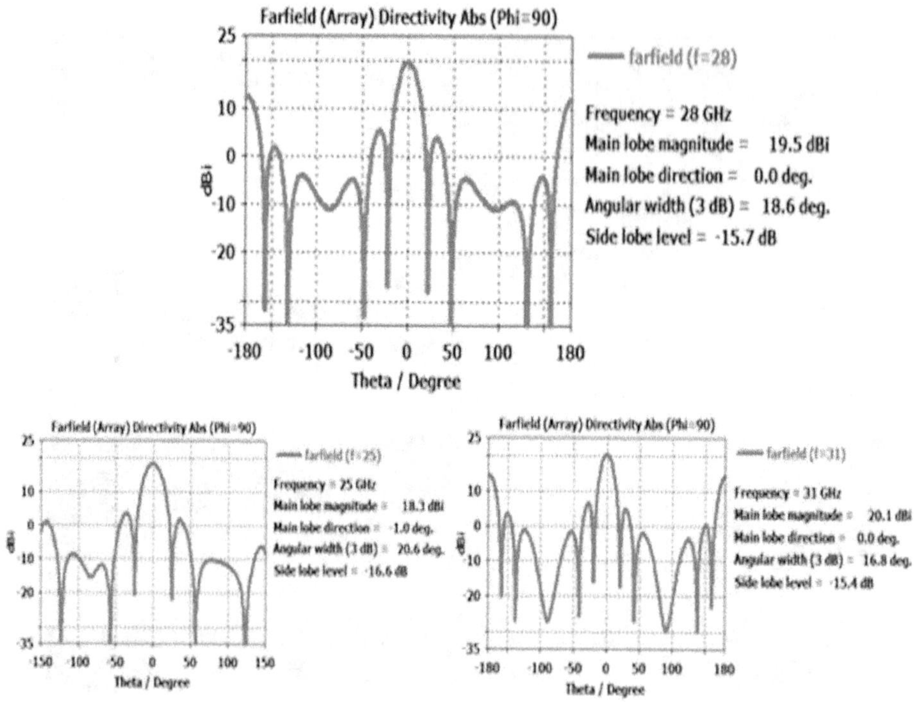

Table 3. Different array sizes with respective gains obtained at the mmWaves spectrum

Array Matrix	Array Size	Array Gain @ 28 GHz
5x5	31mm x 31 mm x 2mm	19.2 dBi
9x9	55mm x 55mm x 2mm	24.1 dBi
17x17	103mm x 103mm x 2mm	29.7 dBi
33x33	199mm x 199mm x 199mm	36.2 dBi
65x65	391mm x 391mm x 199mm	42.1 dBi

The integrated block diagram of the RF energy harvesting along with the presented smart IoT sensor design is presented in Figure 15. Once installed, the 'Search' and 'Charge' operations are configured with the electronically steered beams. The base station directs the radiated beam towards the sensor for recharging the battery. The charging RF wave is shown in blue color at the mmWave spectrum whereas the data exchanged between the sensor and the RFID reader is marked with the black color at the UHF frequency band. The proposed integrated system does not need

Figure 14. Beam steering by 17x17 array

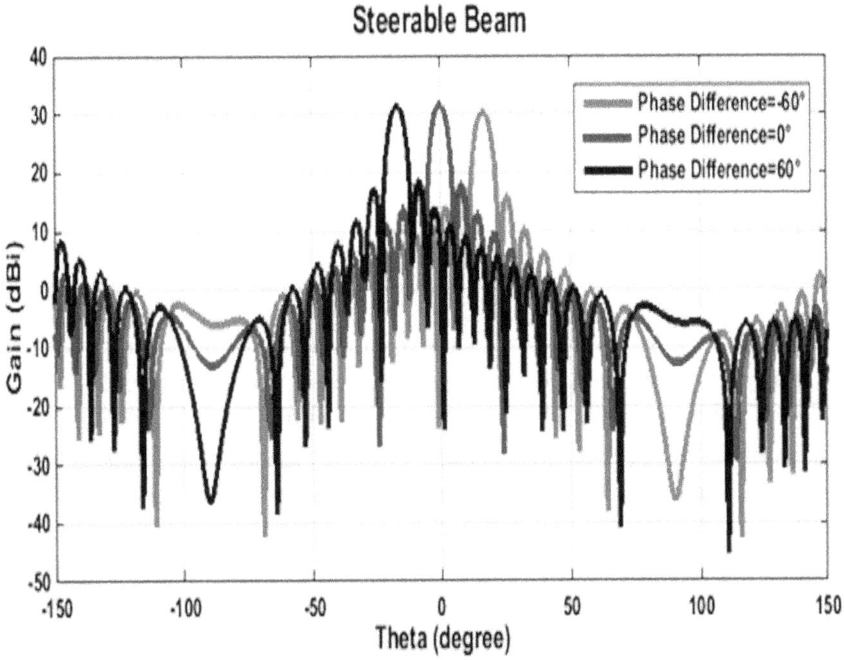

Figure 15. RF energy harvesting scheme for IoT sensor

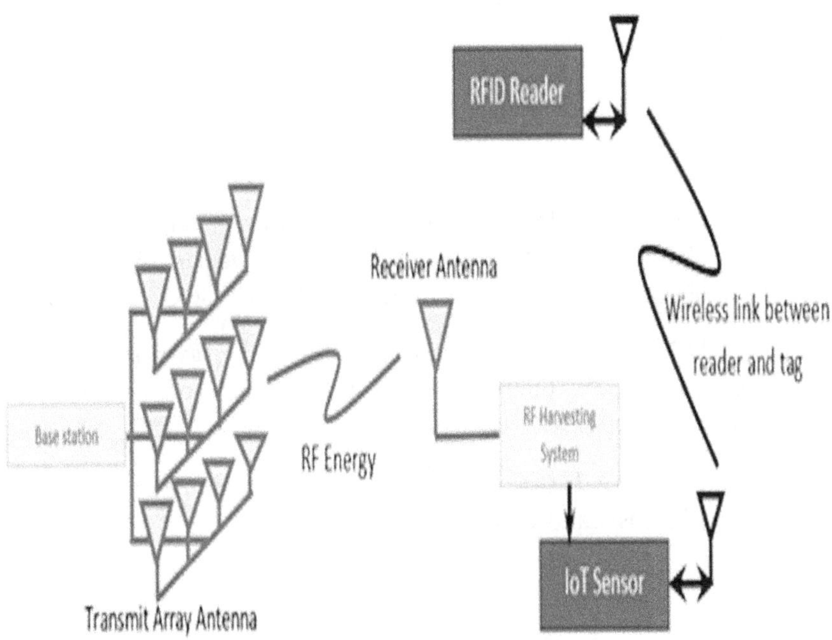

any human intervention and it can be used to charge up any device working over the frequency band 24.25-29.5 GHz (Figure 12) which covers a major portion of the global 5G frequency band (Figure 7).

The charging scheme can be used to power other devices with 5G New Radio mmWaves spectrum configuration.

CONCLUSION

The main purpose of the proposed chapter is to give an innovative way of designing a Connected-Home sensor as an application of IoT services along with the RF Energy Harvesting system. In this regard, the novel sensor design based on the RFID and NFC technologies is explained with functional block diagrams. The presented sensors aid the Connected Home concept with smart controlling systems. To keep such sensors in working condition, the RF Energy Harvesting system is presented in this chapter. The working algorithm of the RFEH system is explained with a design of a font end array antenna at mmWaves, 5G frequency bands. With the implementation of 5G technology, the IoT sensors will have more services for Connected Homes, starting from healthcare to the monitoring of resource distribution. All such initiatives will upgrade the safety and lifestyle of society.

The IoT is destined to have a profound impact, well beyond the home environment. It will transform virtually all industries, from hospitality and retail to automotive, agriculture, and health sector. It is true to mention that the application related to Connected Home is limitless.

REFERENCES

Al-Khayari, A., Al-Khayari, H., Al-Nabhani, S., Bait-Suwailam, M., & Nadir, Z. (2013). *Design of an Enhanced RF Energy Harvesting System for Wireless Sensors*. IEEE GCC Conference And Exhibition. doi:10.1109/IEEEGCC.2013.6705826

Al-Lawati, M., Al-Busaidi, M., & Nadir, Z. (2012). *RF Energy Harvesting System Design for Wireless Sensors*. International Multi-Conference on Systems, Signals, and Devices. doi:10.1109/SSD.2012.6197969

Allee, G. (n.d.). *Smart CITIES USA*. Smart America. Retrieved from https://smartamerica.org/teams/smart-cities-usa/

Cisco. (2016). *Cisco Visual Networking Index Predicts Near-Tripling of IP Traffic by 2020*. Retrieved from https://newsroom.cisco.com/press-release-content?type=press-release&articleId=1771211

D'Amato, G., Avitabile, G., Coviello, G., & Talarico, C. (2016). Towards a Novel Architecture for Beam String of Active Phased-Array Antennas. *International Midwest Symposium on Circuits and Systems (MWSCAS)*.

Felini, C., Merenda, M., & Corte, F. G. D. (2014). Dynamic impedance matching network for RF energy harvesting systems. *IEEE RFID Technology and Applications Conference (RFID-TA)*. 10.1109/RFID-TA.2014.6934206

Gartner. (2015). *Gartner says 6.4 billion Connected "Things" Will Be in Use in 2016, Up 30 Percent From 2015*. Retrieved from https://www.gartner.com/newsroom/id/3165317

Hu, J., Hao, Z. C., & Hong, W. (2017). Design of a Wideband Quad-Polarization Reconfigurable Patch Antenna Array Using a Stacked Structure. *IEEE Transactions on Antennas and Propagation*, *65*(6), 3014–3023. doi:10.1109/TAP.2017.2695529

Khansalee, E., & Nuanyai, K. (2015). High-Frequency Rectifier for RF Energy Harvesting Systems. *International Conference on Information Technology and Electrical Engineering (ICITEE)*. 10.1109/ICITEED.2015.7408961

Kumar, R., Louzir, A., & Naour, J. Y. L. (2016). RFID-based solution for the sensing of home electrical devices activity. *International Symposium on Antenna and Propagation (ISAP2016)*.

Loh, T. H., & Cheadle, D. (2017). An Electronically Steerable Parasitic Array Radiator Antenna with Adaptive Beamforming Ability (A Radiation Pattern Assessment). *International Workshop on Electronics: Applications and Student Innovation Competition (iWEM)*.

Research Nester. (2016). *Internet of things (IoT) Market: Global Demand, Growth Analysis & Opportunity Outlook 2023*. Retrieved from https://www.researchnester.com/reports/internet-of-things-iot-market-global-demand-growth-analysis-opportunity-outlook-2023/216

Statista. (2018). *Internet of Things (IoT) connected devices installed base worldwide from 2015 to 2025 (in billions)*. Retrieved from https://www.statista.com/statistics/471264/iot-number-of-connected-devices-worldwide/

Sun, C., Hirata, A., Ohira, T., & Karmakar, N. C. (2004). Fast Beamforming of Electrically Steerable Parasitic Array Radiator Antennas: Theory and Experiment. *IEEE Transactions on Antennas and Propagation*, *52*(7), 1819–1832. doi:10.1109/TAP.2004.831314

Valdes-Garcia, A., Sadhu, B., Gu, X., Tousi, Y., Liu, D., Scott, K., Hillin, J., Sahl, S., & Rexberg, L. (2018). Circuit and Antenna-in-Package Innovations for Scaled mmWave 5G Phased Array modules. *IEEE Custom Integrated Circuits Conference (CICC)*. 10.1109/CICC.2018.8357050

Waghmare, P., Chandgude, M., Chaure, P., & Chaudhari, A. (2017). Survey on: Home Automation Systems. *International Conference on Trends in Electronics and Informatics (ICEI2017)*.

Wang, L., Wang, J., & Shi, J. (2018). Design of 4x4 Meandering-Fed Stacked Patch Antenna Array. *International Workshop on Antenna Technology (iWAT)*.

Chapter 7
Blockchain Applications and Benefits in Cloud Computing

Ramiz Salama
Near East University, Nicosia, Turkey

Fadi Al-Turjman
AI and Robotics Institute, Near East University, Nicosia, Turkey & Faculty of Engineering, University of Kyrenia, Kyrenia, Turkey

ABSTRACT

Due to its many attributes, including decentralization, transparency, and security, blockchain is now a very important and novel technology for the current generation of industrial applications. One of these fields, the Cloud of Things, was created by fusing cloud computing and the Internet of Things. Blockchain is a key component of this technology because it provides data privacy and network security, eliminates Cloud of Things challenges through decentralization, and incorporates Cloud of Things features through elasticity and scalability functionalities, enhancing the overall efficiency of blockchain operations. As a result, it is highly advantageous to connect blockchain to the Internet of Things (IoT); this technology is known as BCoT, and it is now thought to be a very promising field for a wide range of industrial applications.

INTRODUCTION

Blockchain technology and cloud computing are two transformative innovations that have reshaped the digital landscape. Blockchain, originally designed as a decentralized ledger for cryptocurrencies like Bitcoin, has found applications far beyond digital currencies. Cloud computing, on the other hand, has revolutionized

DOI: 10.4018/979-8-3693-0786-1.ch007

the way businesses store, process, and manage data and applications. When these two technologies converge, they offer a potent combination with numerous benefits (Yu, 2020).

Blockchain Technology Overview

Blockchain is a distributed and immutable ledger that records transactions across a network of computers in a secure and transparent manner. It operates without a central authority, making it resistant to fraud and manipulation. Transactions on a blockchain are grouped into blocks and linked together in a chain. This technology offers several key features (Scarfone, 2019).

Decentralization: No single entity has control over the network, making it tamper-resistant.
Transparency: Transactions are publicly recorded and can be audited by anyone on the network.
Immutability: Once data is recorded on the blockchain, it is nearly impossible to alter or delete.
Security: Strong cryptographic techniques ensure the integrity and privacy of data.

Cloud Computing Overview

Cloud computing is the delivery of on-demand computing resources, such as servers, storage, databases, networking, and software, over the internet. It offers several deployment models, including public, private, hybrid, and multi-cloud, and is characterized by the following features:

Scalability: Resources can be scaled up or down to meet demand, ensuring cost efficiency.
Accessibility: Data and applications can be accessed from anywhere with an internet connection.
Cost Efficiency: Pay-as-you-go pricing models reduce the need for large upfront investments.
Reliability: Data redundancy and failover mechanisms ensure high availability.

Blockchain in Cloud Computing: Applications and Benefits

When blockchain technology is integrated with cloud computing, it opens up a range of applications and benefits:

Enhanced Security: Blockchain's immutability and cryptographic security can be used to secure data stored in the cloud. This is particularly important for sensitive information (Zodpe & Sapkal, 2020).

Data Integrity: By using blockchain to record changes to data stored in the cloud, users can verify the integrity and authenticity of data, ensuring it hasn't been tampered with.

Smart Contracts: Smart contracts, self-executing agreements with the terms of the contract written into code, can automate processes within cloud-based applications, reducing the need for intermediaries.

Supply Chain Management: Blockchain can be used to track and verify the movement of goods and materials within supply chains hosted on cloud platforms.

Identity Management: Combining blockchain and cloud computing can improve identity management systems, enhancing security and reducing identity theft.

Auditing and Compliance: Blockchain's transparency and auditability can simplify compliance with industry regulations and standards in cloud-hosted systems.

Decentralized Applications (DApps): Developers can build decentralized applications on blockchain platforms and host them on cloud infrastructure, creating a more resilient and scalable application environment.

Tokenization: Tokens and cryptocurrencies can be used to facilitate payments and transactions within cloud-based applications.

In conclusion, the integration of blockchain technology with cloud computing offers a compelling fusion of trust, security, and efficiency (Yu, 2020). As these technologies continue to evolve, we can expect even more innovative applications and benefits to emerge, shaping the future of digital infrastructure and data management.

AIM OF THE RESEARCH

Research on blockchain applications and benefits in cloud computing is conducted for several important reasons, as outlined below:

Enhancing Security: Security is a top concern in cloud computing, as data and applications are stored offsite in data centers. Blockchain can provide an additional layer of security through its decentralized and immutable ledger. Research aims to explore how blockchain can be used to secure data, transactions, and access control in cloud environments.

Data Integrity: Data integrity is crucial, especially in cloud environments where multiple users and entities access and modify data. Blockchain technology can be leveraged to ensure data integrity, making it difficult for unauthorized changes or data tampering to occur. Research helps identify how blockchain can maintain data integrity in the cloud.

Smart Contracts and Automation: Smart contracts, which execute automatically when predefined conditions are met, have the potential to streamline various processes in the cloud, such as billing, procurement, and supply chain management. Research investigates the practicality and benefits of using smart contracts in cloud computing.

Efficiency and Cost Reduction: Cloud computing is valued for its cost-efficiency, but blockchain can further optimize operations by reducing the need for intermediaries, improving transparency, and automating tasks. Research helps identify use cases where these efficiencies can be realized.

Supply Chain Management: The integration of blockchain and cloud computing can revolutionize supply chain management, providing transparency, traceability, and security in global supply chains. Research explores how these technologies can be applied to improve supply chain operations.

Identity Management: Identity theft and unauthorized access to cloud resources are common security challenges. Blockchain can provide a secure identity management solution, and research aims to develop and implement effective identity management systems in the cloud.

Auditing and Compliance: Many industries have strict regulatory requirements, and cloud providers need to ensure compliance. Blockchain's transparent and auditable nature can simplify compliance procedures. Research investigates how blockchain can be used to streamline compliance processes.

Decentralized Applications (DApps): Research explores the development of decentralized applications (DApps) that combine the benefits of both blockchain and cloud computing. DApps can offer improved resilience, scalability, and security compared to traditional centralized applications.

Tokenization: The use of tokens and cryptocurrencies in the cloud can simplify payment processes, create new revenue models, and improve financial transparency. Research aims to identify the potential benefits and challenges of tokenization in cloud environments.

Innovation and Competitive Advantage: Organizations that can harness the advantages of blockchain in cloud computing gain a competitive edge. Research enables companies to stay at the forefront of technological advancements and offer innovative solutions to their customers.

Future-Proofing: Research is essential to understand how emerging technologies can be integrated into existing cloud infrastructures, ensuring that businesses can adapt and remain competitive as technology evolves.

In summary, research on blockchain applications and benefits in cloud computing is essential to harness the full potential of these technologies, improve security, efficiency, and transparency, and create innovative solutions that can benefit businesses, consumers, and various industries. This research helps bridge the gap between theory and practical implementation, fostering innovation and progress in the field of cloud computing.

THE NEED FOR BLOCKCHAIN AND COT INTEGRATION

The following are some reasons why conventional CoT networks are ineffective: Traditional CoT confronts a basic challenge in that it relies heavily on centralized communication systems and a single cloud, making it challenging to scale service operations when IoT networks grow significantly Seneviratne, A. (2020). The need for most contemporary CoT systems to outsource IoT data processing to any third party, such as a cloud provider, raises concerns about data privacy, which is the second major challenge that CoT faces. The third major barrier is that centralized network infrastructure increases IoT device power consumption and communication latency, making large-scale CoT deployments in practical applications difficult (Jha, 2022). Given the limitations of CoT and the advantages of Blockchain, fusing the two technologies could potentially address CoT's issues.

CLOUD COMPUTING'S POTENTIAL USE OF BLOCKCHAIN TECHNOLOGY

Some significant areas where blockchain can outperform traditional cloud solutions include the ones listed below:

The blockchain system may be quite helpful in terms of enhancing functionality or performance because of its unique qualities. Because blockchain is a decentralized system, it can help develop an architecture in which numerous computers can work together to accomplish one specific task, such data processing or storage, at once, speeding up data processing and uploading. Another concern is enhancing cloud computing security (Yu, 2020). Data security risks always exist since cloud computing is centralized. For instance, if the system as a whole were compromised, there would be no backup data, and any lost data would be irretrievably destroyed. In order to address these issues, blockchain technology has the potential to be used in cloud computing (Kim & Deka, 2020).

BLOCKCHAIN IN CLOUD COMPUTING BENEFITS

1. Decentralization

The reliance on a centralized server for data management and decision-making presents a fundamental barrier for IoT and cloud computing technologies. For instance, if the central server fails, the entire system may be affected, and important data stored on the server may be lost. The main server is also susceptible to hacker attack (Ishfaq, 2022). In a decentralized system, the likelihood that the entire system will malfunction if one server dies is decreased when many copies of the same data are kept on other computer nodes. Additionally, data loss should be prevented because there are numerous copies of the data stored on different nodes.

2. Enhanced Data Security

IoT stored data is typically associated with the homeowner's personal information, such as video footage, voice recordings, household items, property, and personal habits, and a leak of this data can result in robbery, attacks, and the illegal sale of personal information for money. This makes storing IOT stored data on the cloud a significant challenge. These circumstances lead to a breach of the cloud infrastructure (Almazrooie, 2021). The use of blockchain in cloud computing has the potential to increase overall system security as a response to this problem.

3. Improved Product and Service Ownership Tracing

Keeping track of every vehicle in its network, their present location, the amount of time a vehicle spent at a site, and establishing contact between several cars is one of the logistics industry's most challenging issues (Qureshi, 2022). Software solutions like shipment tracking systems face issues due to their centralized nature. Blockchain holds tremendous potential for monitoring these commodities and services.

4. Tolerance for Faults

The replication of blockchain data across a network of computer systems connected safely via collaborative clouds may benefit from the cloud. This makes it less likely that any cloud node loss will result in a single point of failure, enabling continuous services (Kumari & Kaur, 2021).

Figure 1. Benefits of blockchain on cloud

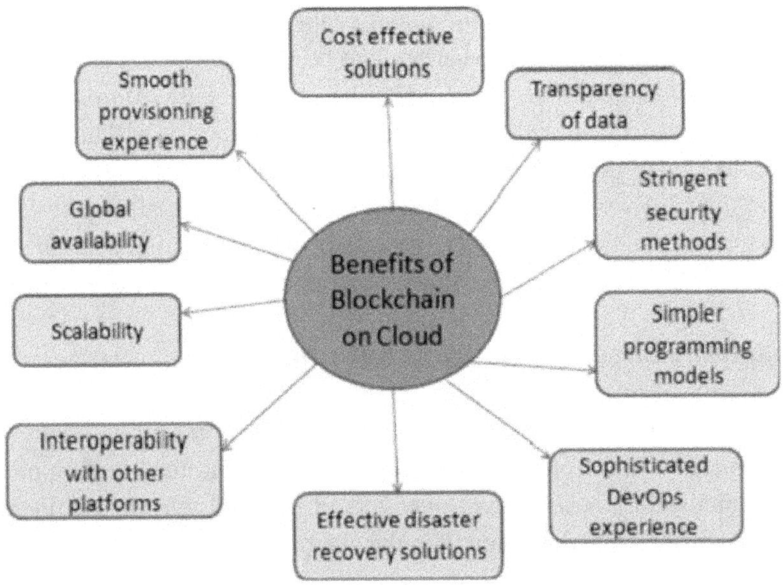

5. Scalability

Large-scale blockchain applications have the potential to significantly increase the number of transactions on blockchain networks. As a result, robust data processing services are needed for scalable blockchain systems with high transaction execution. The scalability of the cloud enables it to offer this market's blockchain activities on-demand processing capacity. By merging blockchain with cloud computing technologies, it is possible to create an integrated system that is highly scalable (Habib, 2022).

COT AND BLOCKCHAIN INTEGRATION APPLICATIONS

1. Healthcare Industry Intelligence

For modernizing and improving current healthcare infrastructure and practices, BCoT has great promise. The healthcare sector is made up of medical facilities and organizations. The mainstays of the industry include healthcare-related services, medical devices and technologies including ventilators, and medical insurance.

Significant service effectiveness and industrial security challenges can be resolved by BCoT. By reducing the amount of time needed for sharing communication from patients' devices to doctors' devices and among connected devices, new smart services, such as CoT enabled Health data sharing, can be provided with the aid of BCoT integration in healthcare (Seneviratne, 2020). Electronic health records (EHRs), for instance, will be processed and kept on the cloud, and doctors will have access to patient medical data via mobile devices (such smartphones) for health monitoring. Blockchain may prove to be especially successful in addressing security issues in the process of exchanging health data because it makes use of decentralized data verification of all peer nodes and message validation through a consensus technique.

2. Blockchain-Based Smart Home Automation

Smart home automation is one of the most significant industries where BCoT might be applied. A house can become a smart home by automating its appliances, providing comfort for its occupants. A network of Internet of Things (IoT) devices that collect environmental data, store it in the cloud, process it, and then carry out a particular action depending on the processed data constitutes a smart home. When a fire is detected, a message is sent to the homeowner and the house's water sprinkler system or fire alarm is immediately activated. A fire detection system, for instance, comprises of a temperature measuring sensor that determines whether or not there is a fire in the home. Blockchain automation could be applied to smart homes to increase security by lowering the possibility of data loss or privacy invasion. A decentralized data integrity architecture that uses blockchain can be used to guarantee the stability and security of the system without the requirement for a third party (Mohanty, 2021).

3. Integrated Blockchain for Smart Transportation

In the contemporary world, the transportation sector has a lot of space for growth. As sensor, computer, and communication technology have expanded quickly in recent years, so have transportation systems. This has had a significant impact on people's lives, including commute times, road safety, and freight transit times. Smart transportation is an Internet of Things (IoT) application pertaining to included communication technology and vehicle services in transportation networks (Seneviratne, 2020). However, dynamic vehicle-to-vehicle communication in automotive situations and an over-reliance on centralized network authority have given birth to security flaws. Blockchain can help create a secure, reliable, and decentralized ITS system in this scenario.

4. Intelligent Manufacturing

A new industry that could profit from BCoT is smart manufacturing. Smart manufacturing is the use of automated devices that can perform particular tasks far more intelligently than they now can. This sector makes use of service-oriented manufacturing, cloud manufacturing, and IoT-enabled technology. However, in the existing environment, problems like centralized industrial networks and third-party authority hinder smart manufacturing (Fu, 2020). The centralized industrial designs lose some of their flexibility, effectiveness, and security. These issues might be resolved by utilizing BCoT, which can improve security while still carrying out the decentralized concept.

3. Issues With the Current Cloud Computing System

According to a Gartner survey of 110 senior executives who commented on the difficulties their various organizations are facing, cloud computing is at the top of the list of rising hazards. How to prevent unauthorized access to a platform with such a huge audience was the main problem that attracted attention. Due to the complexity of cloud computing, fraudsters from all over the world are searching for viruses and system defects that can steal sensitive and personal data. Network failures and a failure to provide access to information may also occur from a cloud provider's inability to function as a result of a disruption. Many free service providers now provide the advantage of storing databases and information. However, some websites take an unreliable position by eliminating content they find offensive without providing customers with sufficient justification. The likelihood of data loss and control dispersion thus rises. An organization's legal requirements might not be satisfied by cloud computing, claims renowned international cryptographer Burce Schneier. The complexity of cloud-based centralized computing and the fast evolution of the regulatory architecture make it more challenging to establish trust and obtain authorization. Although cloud-based businesses have been successful in large firms, small businesses have not been drawn to them because of their high cost and feasibility, which is another depressing trend. The consolidation of cloud services is one of the biggest upheavals, say academics and experts. Data must travel through tightly regulated central servers, which raises costs and reduces efficiency and connectivity. When repositories are centralized, issues with privacy, security, and assessing verifiability arise.

SOLUTION

These servers and repositories may be decentralized in the future. And using Blockchain is undoubtedly the best way to introduce the benefits of devolution to cloud computing. The need for a decentralized network in modernizing cloud computing architecture has been acknowledged by developers all around the world. The IoT market could gain a lot from the integration of distributed ledger technology and artificial intelligence. The idea of edge computing is comparable. In edge computing, some data is displaced from one or more central nodes to end-users or other edge local cloud, grid, fog computing, distributed data storage, and other more advanced solutions. It is a hybrid of peer-to-peer (P2P), ad hoc networking, local cloud, grid, fog computing, distributed data storage, and other more advanced solutions.

What Solutions Would Blockchain Offer for These Issues?

Because it is built on a distributed ledger system, the blockchain cannot be created without the consent of all stakeholders. Blockchains also have the advantage of being immutable, meaning that once data is placed on one, it is there for good. As a result, you can ensure the security of your assets and data kept in the cloud and know where, when, and how your data was used. Therefore, we can all agree that blockchain addresses some, if not all, of the key issues with cloud computing. Blockchain technology has already shown its value with Bitcoin in the world of financial cryptocurrencies, prompting interest from a wide range of other businesses. Therefore, cloud computing will soon imitate the financial sector's example. Additionally, blockchain technology with its incentive system can be useful. The hosts of different nodes can be compensated with tokens, which serve as the network's currency. It's a successful strategy for encouraging interested people to pool their resources and discourage dishonest individuals. With so many sugar cubes, there are some sour places. The institution's integrity will be questioned if a payment system is not in place to give stakeholders a sufficient incentive structure. It might also prevent the distribution of unfair incentives. Scalability, another attribute of cloud computing, is its capacity to handle enormous data volumes. How will Blockchain be able to gather enormous amounts of data, show scalability, and guarantee the effectiveness and speed of big transactions and data transfers given the existing infrastructure systems in place and their flaws? One of the most important characteristics is that the program is properly computed, which makes it less vulnerable to attacks from cybercriminals. It is the worst kind of failure, leading to a serious breach in the confidentiality of sensitive data that, if tampered with, might be exploited to devastating effect. This necessitates placing a focus on improving its security and accuracy in order to remove discrepancies. If a supplier cannot ensure the protection of these fundamental rights, they inadvertently withdraw their offer to provide the services. It's crucial to develop a trustworthy interface that links smart contracts to actual events because a failure in this system could result in a catastrophic calamity. Erasure control is another issue that needs to be handled if and when the wall falls. Regulating the amount of data that is deleted, determining what data should be preserved, and, in some cases, keeping a reliable backup. These abilities also need to be distributed well in order to reduce significant losses.

CONCLUSION

Blockchain's integration with cloud computing offers numerous benefits, such as enhanced data security, decentralization, smart contract automation, and improved trust. However, it's important to note that blockchain technology is not a one-size-fits-all solution, and its adoption in cloud computing should be carefully considered. Some key considerations include Scalability: Blockchain's scalability issues may impact its suitability for large-scale cloud applications. Solutions like sharding and layer 2 technologies are being developed to address this concern. Regulatory Compliance: Blockchain's decentralized nature can pose challenges in complying with data protection and privacy regulations. Solutions should be developed to balance security and compliance. Integration Challenges: Integrating blockchain with existing cloud infrastructure can be complex and may require substantial changes to existing systems. Energy Consumption: Proof-of-work blockchains can be energy-intensive, which can be a concern for environmentally conscious cloud service providers. In conclusion, while blockchain offers promising benefits in cloud computing, its adoption should be strategic and well-planned. It is essential to assess the specific use case, consider the technical challenges, and address regulatory and environmental concerns to fully leverage its advantages.

REFERENCES

Agi, M. A., & Jha, A. K. (2022). Blockchain technology in the supply chain: An integrated theoretical perspective of organizational adoption. *International Journal of Production Economics*, *247*, 108458. doi:10.1016/j.ijpe.2022.108458

Ahmed, N., Michelin, R. A., Xue, W., Ruj, S., Malaney, R., Kanhere, S. S., Seneviratne, A., Hu, W., Janicke, H., & Jha, S. K. (2020). A survey of COVID-19 contact tracing apps. *IEEE Access : Practical Innovations, Open Solutions*, *8*, 134577–134601. doi:10.1109/ACCESS.2020.3010226

Awadallah, R., Samsudin, A., Teh, J. S., & Almazrooie, M. (2021). An integrated architecture for maintaining security in cloud computing based on blockchain. *IEEE Access : Practical Innovations, Open Solutions*, *9*, 69513–69526. doi:10.1109/ACCESS.2021.3077123

Dreyfus, J., Audureau, E., Bohbot, Y., Coisne, A., Lavie-Badie, Y., Bouchery, M., Flagiello, M., Bazire, B., Eggenspieler, F., Viau, F., Riant, E., Mbaki, Y., Eyharts, D., Senage, T., Modine, T., Nicol, M., Doguet, F., Nguyen, V., Le Tourneau, T., ... Messika-Zeitoun, D. (2022). TRI-SCORE: A new risk score for in-hospital mortality prediction after isolated tricuspid valve surgery. *European Heart Journal*, *43*(7), 654–662. doi:10.1093/eurheartj/ehab679 PMID:34586392

Fu, L., Wang, B., Yuan, T., Chen, X., Ao, Y., Fitzpatrick, T., Li, P., Zhou, Y., Lin, Y., Duan, Q., Luo, G., Fan, S., Lu, Y., Feng, A., Zhan, Y., Liang, B., Cai, W., Zhang, L., Du, X., ... Zou, H. (2020). Clinical characteristics of coronavirus disease 2019 (COVID-19) in China: A systematic review and meta-analysis. *The Journal of Infection*, *80*(6), 656–665. doi:10.1016/j.jinf.2020.03.041 PMID:32283155

Hsu, C. Y., Yang, W., Parikh, R. V., Anderson, A. H., Chen, T. K., Cohen, D. L., He, J., Mohanty, M. J., Lash, J. P., Mills, K. T., Muiru, A. N., Parsa, A., Saunders, M. R., Shafi, T., Townsend, R. R., Waikar, S. S., Wang, J., Wolf, M., Tan, T. C., ... Go, A. S. (2021). Race, genetic ancestry, and estimating kidney function in CKD. *The New England Journal of Medicine*, *385*(19), 1750–1760. doi:10.1056/NEJMoa2103753 PMID:34554660

Iqbal, T., & Qureshi, S. (2022). The survey: Text generation models in deep learning. *Journal of King Saud University. Computer and Information Sciences*, *34*(6), 2515–2528. doi:10.1016/j.jksuci.2020.04.001

Kumar, N., Mishra, D., Kim, S. Y., & Jin, S. H. (2020). Self-assembled nis-sns heterostructure via facile successive adsorption and reaction method for high-performance solid-state asymmetric supercapacitors. *Thin Solid Films*, *709*, 138138. doi:10.1016/j.tsf.2020.138138

Kumari, P., & Kaur, P. (2021). A survey of fault tolerance in cloud computing. *Journal of King Saud University. Computer and Information Sciences*, *33*(10), 1159–1176. doi:10.1016/j.jksuci.2018.09.021

Liu, L., Gudmundsson, L., Hauser, M., Qin, D., Li, S., & Seneviratne, S. I. (2020). Soil moisture dominates dryness stress on ecosystem production globally. *Nature Communications*, *11*(1), 4892. doi:10.1038/s41467-020-18631-1 PMID:32994398

Yaga, D., Mell, P., Roby, N., & Scarfone, K. (2019). *Blockchain technology overview*. arXiv preprint arXiv:1906.11078.

Zhang, X., Tan, Y., Ling, Y., Lu, G., Liu, F., Yi, Z., Jia, X., Wu, M., Shi, B., Xu, S., Chen, J., Wang, W., Chen, B., Jiang, L., Yu, S., Lu, J., Wang, J., Xu, M., Yuan, Z., ... Lu, H. (2020). Viral and host factors related to the clinical outcome of COVID-19. *Nature*, *583*(7816), 437–440. doi:10.1038/s41586-020-2355-0 PMID:32434211

Zodpe, H., & Sapkal, A. (2020). An efficient AES implementation using FPGA with enhanced security features. *Journal of King Saud University. Engineering Sciences*, *32*(2), 115–122. doi:10.1016/j.jksues.2018.07.002

Zulfiqar, U., Jiang, W., Xiukang, W., Hussain, S., Ahmad, M., Maqsood, M. F., Ali, N., Ishfaq, M., Kaleem, M., Haider, F. U., Farooq, N., Naveed, M., Kucerik, J., Brtnicky, M., & Mustafa, A. (2022). Cadmium phytotoxicity, tolerance, and advanced remediation approaches in agricultural soils; a comprehensive review. *Frontiers in Plant Science*, *13*, 773815. doi:10.3389/fpls.2022.773815 PMID:35371142

Chapter 8
Harnessing the Power of Machine Learning for Parkinson's Disease Detection

Neepa Biswas
https://orcid.org/0000-0003-2790-1768
Narula Institute of Technology, India

Bannishikha Banerjee
https://orcid.org/0000-0003-4848-8749
Amity University, Kolkata, India

Debarpita Santra
https://orcid.org/0000-0001-5651-5642
Amity University, Kolkata, India

Sudarsan Biswas
RCC Institute of Information Technology, India

ABSTRACT

Parkinson's disease (PD) is a progressive neurodegenerative disorder affecting millions worldwide. Early detection of PD is crucial for effective treatment and management of the disease. Deep learning (DL) and machine learning (ML) have emerged as promising approaches for detecting PD. In this study, a comparative performance analysis is done for DL and ML applications based on speech signals. DL methods using convolutional neural networks (CNNs), long short-term memory (LSTM) networks, and ML methods employing random forest and the XGBoost model were trained and assessed. Performance of the models are evaluated using a variety of performance metrics, including accuracy, precision, recall, and F1-score. Results showed that the XGBoost model outperformed the DL models in terms of accuracy and F1 score, while the CNN and LSTM models achieved higher precision and recall. These findings suggest that XGBoost can be a useful tool for detecting PD based on speech signals, particularly in scenarios where interpretability and computational efficiency are important.

DOI: 10.4018/979-8-3693-0786-1.ch008

1. INTRODUCTION

Parkinson's disease (PD) is becoming an important degenerative disease of the central nervous system, affecting the quality of life of millions of seniors worldwide (Rana et al., 2022). Due to the variability of the disease, PD symptoms might proceed differently from one individual to the next. Patients with Parkin- son's disease may have symptoms including tremors, particularly while at rest. There are several tremor kinds that can occur, including hand tremors, limb stiffness, and gait and balance issues. Generally speaking, movement-related (also known as motor) and non-movement-related (also known as non-motor) symptoms of PD may be separated. Patients showing non-motor symptoms are more affected than those whose main symptoms are motor (Poirier et al., 2016). Non-motor symptoms may include depression, anxiety, sleep behavior disorders, loss of sense of smell, and cognitive impairment (Poirier et al., 2016). So, early detection of PD is crucial for effective treatment and management of the dis- ease. Machine learning has emerged as a promising approach for detecting PD based on various features extracted from different modalities such as handwrit- ing, speech, gait, optical coherence tomography (OCT), and neuroimaging (EEG, MRI, and PET) (Mei et al., 2021). Parkinson's disease begins with very mod- est and perhaps unnoticeable early symptoms, but as the disease advances, the symptoms worsen. Around 90% of Parkinson's patients have voice dysfunction issues, which is a sign of the disease at an early stage, according to research (Mittal and Sharma, 2021).

Machine learning algorithms can learn patterns and relationships in data to develop accurate predictive models. These models can be trained on large datasets of clinical and biomedical data to identify patterns that are indicative of PD. The goal of detecting PD using machine learning is to develop accurate and reliable models that can classify individuals as either having PD or being healthy based on their features. In recent years, XGBoost has emerged as a pop- ular machine learning algorithm for detecting PD due to its ability to handle high-dimensional, sparse data and its strong predictive performance. XGBoost is an ensemble learning algorithm that combines multiple weak learners (decision trees) to improve the overall accuracy of the model (Ferreira et al., 2022). Overall, detecting Parkinson's disease using machine learning has the potential to improve early diagnosis and treatment of the disease. It can also provide in- sights into the underlying mechanisms of PD and help develop new therapies and interventions to manage the disease (Makarious et al., 2022).

Unfortunately, there are not enough skilled medical professionals in developing nations like India and other places. Therefore, identifying or diagnosing Parkinson's is a difficult undertaking because professionals are stressed out by their heavy workload. This inspired us to create a decision assistance system that would help doctors diagnose Parkinson's disease. Use of machine learning can provide a second opinion in the diagnosis of Parkinson's disease and lower the risk of error. In this work, various machine-learning approaches have been applied to detect PD, including Random Forest, CNN, LSTM and XGBoost. These methods have shown promising results in detecting PD based on speech signals.

The overall organization of the study contains Related work in Section 2, Problem Statement at Section 3 Methodology of the proposed work as Section 4, Dataset details at Section 5, Classification model discussion at Section 6 Results and analysis at Section 7 and Finally conclusion at Section 8.

2. RELATED WORKS

There have been various studies conducted to deal with the early detection of Parkinson's disease using XGBoost, Deep Learning and various machine learning models. These work demonstrate the effectiveness of XGBoost in detect- ing Parkinson's disease at an early stage using voice samples. The results show that XGBoost can achieve high accuracy, sensitivity, and specificity in detecting Parkinson's disease, which can potentially improve early diagnosis and treatment of the disease. Parkinson's disease can be detected using deep neural networks such as CNN (Johri et al., 2019). However, as shown in our finding it is observed that the accuracy of Parkinson's' disease detection using CNN has low precision. In neurological diseases such as dementia, Alzheimer and Parkinson, the dis- ease can be detected based on feature speech signals therefore by performing

Support Vector Machine (SVM) algorithms the detection of Parkinson's' dis- ease is performed by Soumaya et al. (2021). However, it has been observed that Support Vector Machine (SVM) is not suitable when there is a large dataset that is being used and produces a lot of noise, datasets such as the one that has been used for this research purpose. Parkinson's patients can be diagnosed using voice dysfunction patterns, according to Lahmiri and Shmuel (2019). A nonlinear SVM classifier and eight distinct pattern ranking methods for feature selection were employed to differentiate between Parkinson's disease patients and healthy people. The Wilcox statistic feature rank approach had an accuracy of 92.21%. A model for detecting Parkinson's disease using vowels was reported by Yaman et al. (2020). The ReliefF method was used to identify the dataset's acoustic features, and KNN and SVM classifiers were used to classify them.

Random Forest classifier and Principal Component Analysis (PCA) have been used to point out patients who are healthy and patients who have Parkin- son's by Xu and Pan (2020). It has been observed on the handwritten dynamics dataset. While this research primarily focuses on the speech impairment of those affected. While performing the comparative study it has also been observed that the algorithm is slow and has a low precision rate for real-time prediction. A hybrid model for Parkinson's diagnosis using speech signals were put up by Po- lat (2019). The SMOTE method was used to initially preprocess the dataset. Random forest classifier was used to accomplish the classification, and 94.8% accuracy was attained.

In order to categorize the model and identify Parkinson's Disease in its early stages, CNN and XGBoost have been chosen in article (Kumar et al., 2022). S et al. (2022) evaluated the effectiveness of classifiers that performed well in terms of accuracy, including Decision trees with ensemble classifier models like XGBoost and Random Forest, which they used in conjunction with various ma- chine learning models.

Long Short-Term Memory has also been used to detect Parkinson's based on non-motor function. Through the use of voice samples from subjects, Rizvi et al. (2020) investigate the potential of a deep neural network (DNN) and long short- term memory (LSTM) network-based model for Parkinson's disease prediction. On the dataset, numerous simulations were run in order to demonstrate the effectiveness of the models and to compare them to traditional machine learning methods. The outcomes demonstrate accuracy for DNN and LSTM of 97.12% and 99.03%, respectively. For the detection of Freezing of Gait (FOG), a deep learning model called the Long Short Term Memory (LSTM) network-based patient-dependent model was used by Ashour et al. (2020). The signals from the three sensors were used to compare the proposed model to more established machine learning techniques, such as the linear support vector machine (SVM). The outcomes proved that the LSTM model, which outperformed the SVM by a margin of average accuracy of 83.38% to 79.48%, was superior.

However, as shown in our comparative study Long Short-Term Memory un- derperforms as compared to XGBoost and has a low precision and accuracy as compared to others, and it the lowest performing algorithm.

3. PROBLEM STATEMENT

Millions of individuals throughout the world suffer from the progressive neuro- logical ailment known as Parkinson's disease (PD). The pathological hallmark of PD is the accumulation of α -Synuclein protein in the brain. Machine learning algorithms have been used to detect PD from various sources of data, such as voice recordings, gait analysis, and brain scans. Techniques, including XGBoost and deep learning, have shown great potential for detecting Parkinson's disease using various types of data, such as speech signals, brain imaging, and clinical assessments. XGBoost is a machine learning algorithm that uses decision trees to model and predict outcomes, while deep learning uses neural networks to learn and represent data. The problem statement for this paper is to compare the performance of XGBoost, deep learning, and other machine learning models for detecting Parkinson's disease using multiple sources of data and to determine the most accurate and efficient method for early detection of Parkinson's disease. This paper will involve collecting and preprocessing data from various sources, training and testing different machine learning models, and evaluating the perfor- mance of each model using several criteria, including F1-score, recall, accuracy, and precision. The technical aspect of this paper will involve implementing and optimizing machine learning models using Python and various libraries, such as sci-kit-learn. Additionally, data visualization and statistical analysis will be con- ducted using tools like Matplotlib and Pandas. The results of this project will provide valuable insights into the most effective and efficient machine-learning approach for detecting Parkinson's disease.

4. METHODOLOGY

This section covers the methods used for the study, including data collection, data pre-processing, and discussion of the several classification models employed for our proposal on the categorization of Parkinson's disease represented in Fig- ure 1. The methodology of this project involves the following stages.

Data acquisition and preprocessing: We acquired and preprocessed var- ious types of data, including speech signals, brain imaging, and clinical assess- ments. This may involve applying preprocessing techniques, such as data clean- ing, normalization, and feature extraction.

Feature selection and engineering: We identified the most relevant fea- tures or patterns that are indicative of Parkinson's disease. This involves select- ing or engineering features that have high discriminative power using statistical methods, machine learning algorithms, or domain knowledge.

Model selection and training: The suitable machine learning must be chosen after the necessary characteristics have been chosen to train and test. This may involve using Machine learning approaches like XGBoost, Random Forest, and Deep Learning models, including Convolutional Neural Networks (CNNs) and Recurrent Neural Networks (RNNs) specifically Long short-term memory networks (LSTM) model, and tuning hyperparameters to improve the accuracy and efficiency of the models.

Figure 1. Methodology for the proposed work

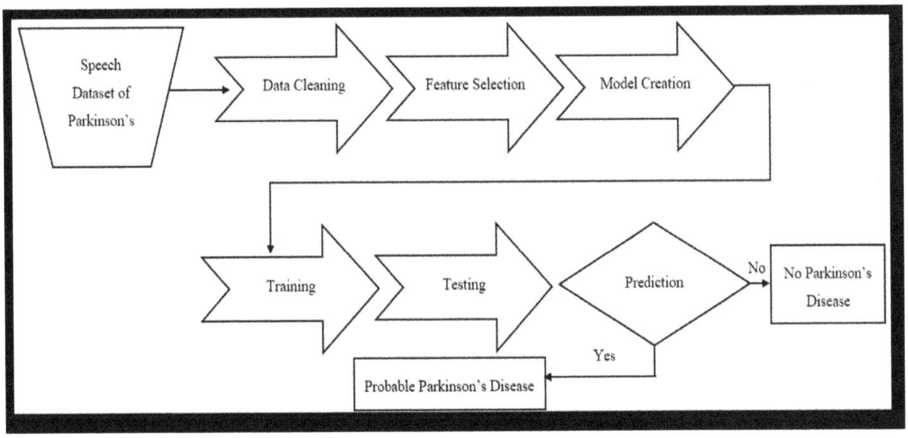

Evaluation and validation: Once the models were trained, they were eval- uated and validated. This involved assessing the effectiveness of the models as measured by common measures like F1-score, recall, accuracy, and precision.

Deployment and scalability: Finally, the developed models need to be deployed in a scalable and practical manner, where they can be used by clinicians and healthcare professionals to detect Parkinson's disease.

Overall, the scope of using Machine learning and Deep Learning for detecting Parkinson's disease is interdisciplinary and involves expertise in various areas, including machine learning, signal processing, neuroscience, and clinical practice.

5. DATA ACQUISITION

For our study, we used the publicly available Oxford Parkinson's Disease Telemonitoring Dataset (Tsanas et al., 2009). This dataset consists of 42 early-stage Parkinson's disease patients who participated in a six-month experiment with telemonitoring equipment for remote symptom progression surveillance and provided a variety of biological voice measurements. The voice recordings were automatically captured from the patient's homes.

The columns of the table include subject number, subject gender, subject age, time since baseline enrollment date, motor UPDRS, total UPDRS, and 16 biomedical voice measures. Attribute information is explained at Table 5.1. One of the 5,875 voice recordings from these people is represented by each row.

Table 1. Attributes of Parkinson's voice dataset

Attribute	Description
Subject	Unique identification number for each subject
age	Age of Subject
sex	Gender of Subject '0' =male, '1' = female
test time	Time since the trial's recruitment.
motor UPDRS **total UPDRS**	Clinician's linearly interpolated motor UPDRS score
Jitter(%),Jitter(Abs), Jitter:RAP,Jitter: PPQ5,Jitter:DDP	Various measures of fundamental frequency fluctuation.
Shimmer,Shimmer(dB), Shimmer:APQ3, Shimmer:APQ5, Shimmer:APQ11, Shimmer:DDA	Several measures of amplitude variation
NHR,HNR	Two measurements of the voice's noise tonal component ratio
RPDE	A measure of nonlinear dynamic complexity
DFA	Exponent of signal fractal scaling
PPE	A basic frequency fluctuation nonlinear measure

The data is in ASCII CSV format. Each row of the CSV file contains a single instance for each voice recording. About 200 recordings are available for each patient, whose subject number is displayed in the first column.

6. CLASSIFICATION TECHNIQUE

Long Short-Term Memory: Long short-term memory networks, or LSTMs, are employed in Deep learning. Many recurrent neural networks (RNNs) are able to learn long-term dependencies, particularly in tasks involving sequence prediction. With the exception of singular data points, like photographs, the LSTM contains feedback connections, meaning it can analyze the whole sequence of data (Sherstinsky, 2020). This has uses in machine translation and speech recog- nition, among others. A unique version of RNN called LSTM exhibits exceptional performance on a wide range of issues.

XGBoost: XGBoost is a distributed gradient boosting library that has been optimized for quick and scalable machine learning model training. A number of weak models' predictions are combined using this ensemble learning technique to get a stronger prediction (Chen and Guestrin, 2016). Extreme Gradient Boosting, or XGBoost, is one of the most well-known and widely used machine learning algorithms because it can handle large datasets and perform at the cutting edge in many machine learning tasks like regression and segmentation.

Random Forest: Due to its excellent accuracy, resilience, feature signifi- cance, adaptability, and scalability, Random Forest is a well-known machine learning technique used for classification and regression problems (Segal, 2004). By averaging many decision trees, Random Forest lessens overfitting and is less susceptible to noise and outliers in the data.

Convolutional Neural Network: A deep learning network design that learns directly from data is a convolutional neural network (CNN or ConvNet) (Albawi et al., 2017). CNNs are very helpful for recognizing objects, classifica- tions, and categories in photos by looking for patterns in the images. They may be quite useful for categorizing signals, time-series, and audio data.

7. RESULTS AND ANALYSIS

In the proposed work, The training and testing of the models are done in Jupyter Notebook – Anaconda. The accuracy, sensitivity, specificity, F1 score, and area under the receiver operating characteristic (AUC) curve are common assessment metrics used to assess the efficacy of machine learning approaches in the identi- fication of Parkinson's disease. These metrics are used to determine the overall performance of the deep learning models in classifying patients with Parkinson's disease correctly. The models used in this research are Random Forest, XGBoost, Convolutional Neural Networks (CNN), and Long Short-Term Memory (LSTM). The Oxford University Parkinson's dataset (Tsanas et al., 2009) is used for the proposed work.

The dataset included voice data, which was cleaned and converted to CSV. Later on, we distinguished the status of the disease employing the following bar chart. The following Figure 2 indicates the presence of Parkinson's disease whereas 0 indicates that the patient does not have Parkinson's.

The authors trained the system using the machine learning models like Ran- dom Forest, XGBoost, CNN and LSTM. The results obtained are diagrammat- ically explained in the following figures. Figure 3 shows the confusion matrix of the actual vs the predicted data by the proposed approach using Random Forest. The following Figure 4 shows the false positive rate, determining the efficiency of the Random Forest model. Figure 5 shows the confusion matrix of the actual vs the predicted data by the proposed approach using XGBoost. The following Figure 6 shows the false positive rate, determining the efficiency of the XGBoost model. Figure 7 shows the confusion matrix of the actual vs the predicted data by the proposed approach

Figure 2. Presence of Parkinson's disease

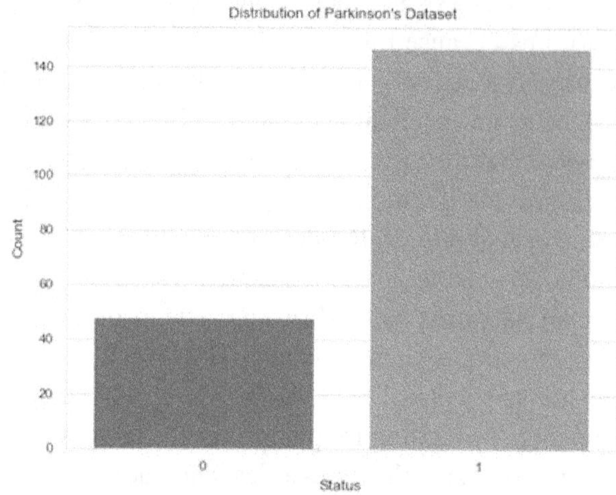

Figure 3. Confusion matrix using random forest

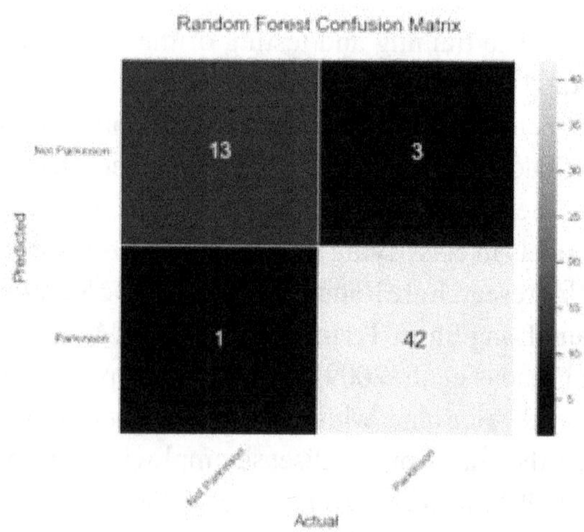

using LSTM. The following Figure 8 shows the false positive rate, determining the efficiency of the LSTM model. Figure 9 shows the confusion matrix of the actual vs the predicted data by the proposed approach using CNN. The following Figure 10 shows the false positive rate, determining the efficiency of the CNN model. Finally, the authors plotted the performance of all the models in the following bar chart. Figure 11 shows the comparative analysis of all the models.

Figure 4. False positive rate using random forest

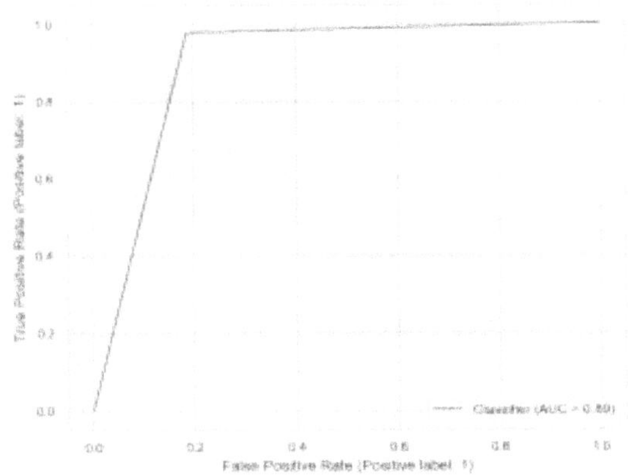

Figure 5. Confusion matrix using XG-boost

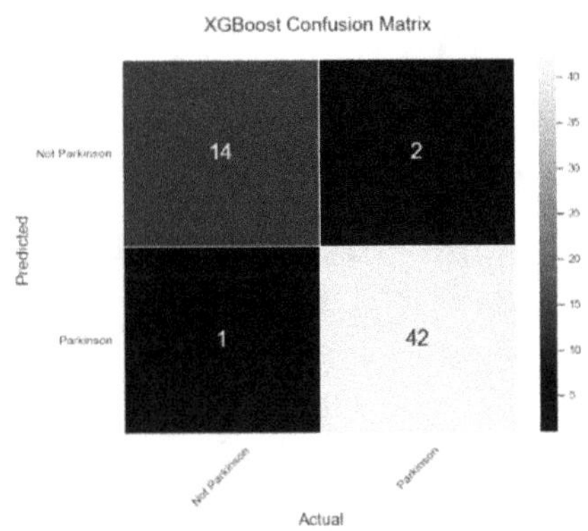

Figure 6. False positive rate using XG-boost

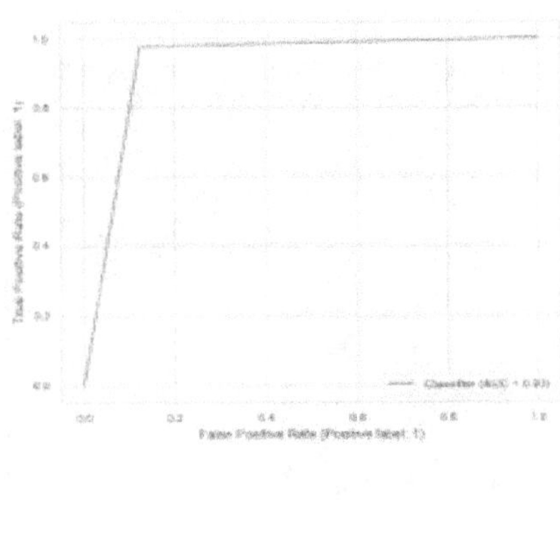

Figure 7. Confusion matrix using LSTM

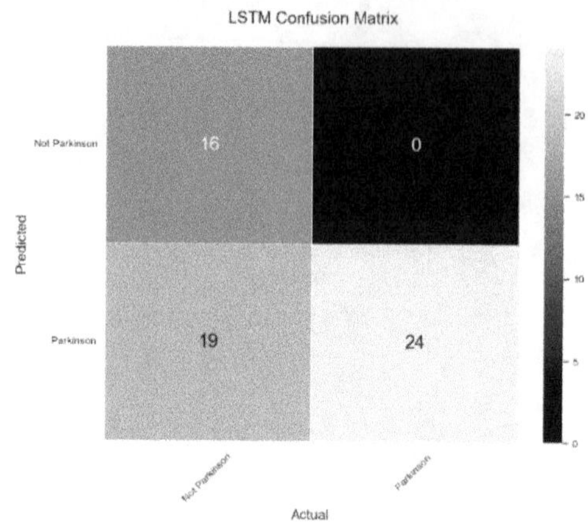

Figure 8. False positive rate using LSTM

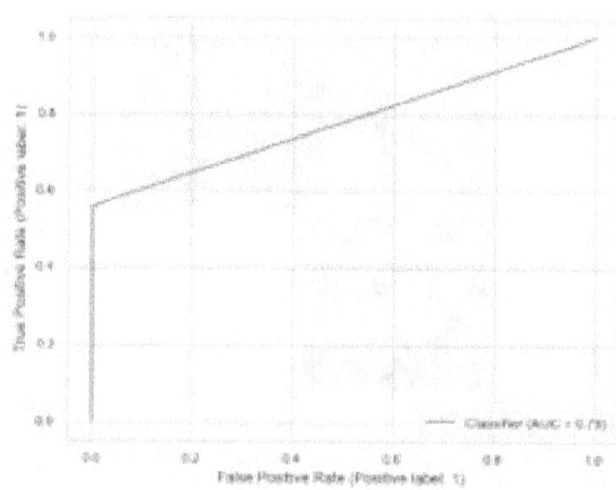

Figure 9. Confusion matrix using CNN

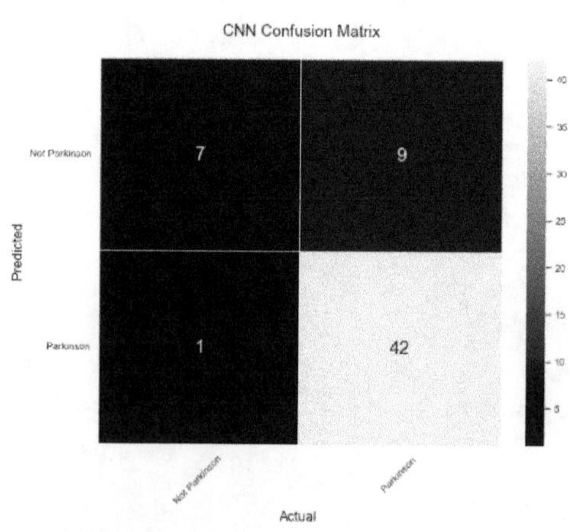

Figure 10. False positive rate using CNN

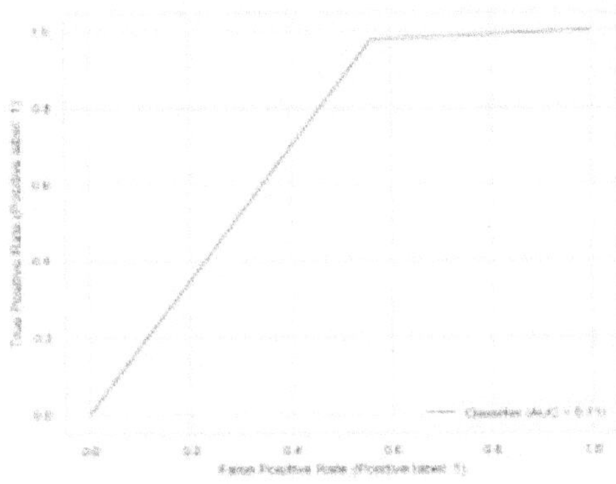

Figure 11. Comparison of training and validation accuracy

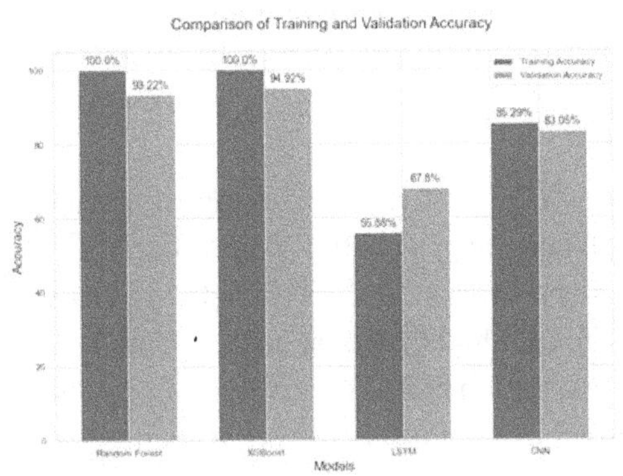

Table 2. Comparison of training and validation accuracy

Model	Accuracy	Sensitivity	Specificity	F1 Score	AUC-ROC
CNN	83.05%	88.89%	93.10%	0.89	0.71
LSTM	72.88%	79.01%	90.56%	0.84	0.50
XGBoost	94.92%	98.40%	94.90%	0.97	0.93
Random Forest	93.22%	94.45%	90.02%	0.95	0.89

In the above graph, it is observed that XGBoost is the best-performing model in comparison to other models which are Random Forest, LSTM, and CNN. It has been observed that the validation accuracy of XGBoost is 94.92% and the training accuracy is 100%. Whereas a lesser performing model such as Random forest has a training accuracy of 100% and validation accuracy of 93.22% respectively CNN has a training accuracy of 86.03% and validation accuracy of 83.05%. Finally, it is observed that LSTM has a training accuracy of 76.47% and validation accuracy of 72.88%. In detail, the result analysis is represented in Table 2.

8. CONCLUSION

XGBoost is a powerful machine learning algorithm that can be used to accurately detect Parkinson's disease using various features from different sources. It offers several advantages such as faster computation time and better interpretability compared to deep learning methods. The evaluation metrics such as accuracy, sensitivity, specificity, F1 score, and AUC-ROC demonstrate the effectiveness of XGBoost in detecting Parkinson's disease. While deep learning methods can achieve higher accuracy, they require large data, computational resources, and careful tuning of hyperparameters. Therefore, XGBoost can be a suitable alter- native for Parkinson's disease detection, particularly in scenarios where limited data or computational resources are available.

Studies have shown XGBoost is a promising technique for detecting Parkin- son's disease with high accuracy, sensitivity, specificity, F1 score, and AUC-ROC score. We compared the performance of XGBoost with CNN, LSTM and Ran- dom Forest algorithms and found that XGBoost outperformed these techniques in terms of accuracy and other evaluation metrics. Furthermore, XGBoost is computationally efficient and can handle large datasets, making it a suitable technique for detecting Parkinson's disease in real-world scenarios.

In the case of detecting the progression of Parkinson's disease, further re- search is needed to validate the effectiveness of this approach and optimize its performance for clinical applications. Additional studies may be required to eval- uate the accuracy and reliability of the models developed using this methodology and to assess the generalizability and scalability of the approach across different populations and clinical settings.

In conclusion, the approach presented using machine learning and deep learn- ing to detect the progression of Parkinson's disease holds great promise for im- proving the diagnosis and treatment of the disease. With further research and development, this approach could potentially transform the field of Parkinson's disease diagnosis and treatment, offering new hope for patients and clinicians alike.

REFERENCES

Albawi, S., Mohammed, T. A., & Al-Zawi, S. (2017). Understanding of a convolutional neural network. 2017 international conference on engineering and technology (ICET), 1–6. doi:10.1109/ICEngTechnol.2017.8308186

Ashour, A. S., El-Attar, A., Dey, N., Abd El-Kader, H., & Abd El-Naby, M. M. (2020). Long short term memory based patient-dependent model for fog detection in parkinson's disease. *Pattern Recognition Letters*, *131*, 23–29. doi:10.1016/j.patrec.2019.11.036

Chen, T., & Guestrin, C. (2016). Xgboost: A scalable tree boosting system. *Proceedings of the 22nd acm sigkdd international conference on knowledge discovery and data mining*, 785–794. 10.1145/2939672.2939785

Ferreira, M. I. A., Barbieri, F. A., Moreno, V. C., Penedo, T., & Tavares, J. M. R. (2022). Machine learning models for parkinson's disease detection and stage classification based on spatial-temporal gait parameters. *Gait & Posture*, *98*, 49–55. doi:10.1016/j.gaitpost.2022.08.014 PMID:36049418

Johri, A., & Tripathi, A. (2019). Parkinson disease detection using deep neural networks. *2019 twelfth international conference on contemporary computing (IC3)*, 1–4.

Kumar, A., Bhatt, B. R., Anitha, P., Yadav, A. K., Devi, K. K., & Joshi, V. C. (2022). A new diagnosis using a parkinson's disease xgboost and cnn-based classification model using ml techniques. *2022 International Conference on Advanced Computing Technologies and Applications (ICACTA)*, 1–6.

Lahmiri, S., & Shmuel, A. (2019). Detection of parkinson's disease based on voice patterns ranking and optimized support vector machine. *Biomedical Signal Processing and Control*, *49*, 427–433. doi:10.1016/j.bspc.2018.08.029

Makarious, M. B., Leonard, H. L., Vitale, D., Iwaki, H., Sargent, L., Dadu, A., Violich, I., Hutchins, E., Saffo, D., & Bandres-Ciga, S. (2022). Multi- modality machine learning predicting parkinson's disease. *NPJ. Parkinson's Disease*, *8*(1), 35. PMID:35365675

Mei, J., Desrosiers, C., & Frasnelli, J. (2021). Machine learning for the diagnosis of parkinson's disease: A review of literature. *Frontiers in Aging Neuroscience*, *13*, 633752. doi:10.3389/fnagi.2021.633752 PMID:34025389

Mittal, V., & Sharma, R. (2021). Classification of parkinson disease based on analysis and synthesis of voice signal. *International Journal of Healthcare Information Systems and Informatics*, *16*(4), 1–22. doi:10.4018/IJHISI.20211001.oa30

Poirier, A.-A., Aub'e, B., C^ot'e, M., Morin, N., Di Paolo, T., & Soulet, D. (2016). Gastrointestinal dysfunctions in parkinson's disease: Symptoms and treatments. *Parkinson's Disease*, *2016*, 2016. doi:10.1155/2016/6762528 PMID:28050310

Polat, K. (2019). A hybrid approach to parkinson disease classification using speech signal: the combination of smote and random forests. 2019 scientific meeting on electrical-electronics & biomedical engineering and computer science (EBBT), 1–3. doi:10.1109/EBBT.2019.8741725

Rana, A., Dumka, A., Singh, R., Panda, M. K., Priyadarshi, N., & Twala, B. (2022). Imperative role of machine learning algorithm for detection of parkinson's disease: Review, challenges and recommendations. *Diagnostics (Basel)*, *12*(8), 2003. doi:10.3390/diagnostics12082003 PMID:36010353

Rizvi, D. R., Nissar, I., Masood, S., Ahmed, M., & Ahmad, F. (2020). An lstm based deep learning model for voice-based detection of parkinson's disease. *Int. J. Adv. Sci. Technol*, *29*(8).

S, E. F., T C, E. S., & R S, V. D. (2022). Prediction of parkinson's disease using xgboost. *2022 8th International Conference on Advanced Computing and Communication Systems (ICACCS)*, *1*, 1769–1772.

Segal, M. R. (2004). *Machine learning benchmarks and random forest regression*. Academic Press.

Sherstinsky, A. (2020). Fundamentals of recurrent neural network (rnn) and long short-term memory (lstm) network. *Physica D. Nonlinear Phenomena*, *404*, 132306.

Soumaya, Z., Taoufiq, B. D., Benayad, N., Yunus, K., & Abdelkrim, A. (2021). The detection of parkinson disease using the genetic algorithm and svm classifier. *Applied Acoustics*, *171*, 107528. doi:10.1016/j.apacoust.2020.107528

TsanasA.LittleM.McSharryP.RamigL. (2009). Accurate telemoni- toring of parkinson's disease progression by non-invasive speech tests. Nature Precedings, 1–1.

Xu, S., & Pan, Z. (2020). A novel ensemble of random forest for assisting diagnosis of parkinson's disease on small handwritten dynamics dataset. *International Journal of Medical Informatics*, *144*, 104283. doi:10.1016/j.ijmedinf.2020.104283 PMID:33010729

Yaman, O., Ertam, F., & Tuncer, T. (2020). Automated parkinson's disease recognition based on statistical pooling method using acoustic features. *Medical Hypotheses*, *135*, 109483. doi:10.1016/j.mehy.2019.109483 PMID:31954340

KEY TERMS AND DEFINITIONS

Convolutional Neural Networks: It is a deep learning network design that derives its knowledge directly from data. CNNs are very helpful for recognizing objects, classifications, and categories in photos by looking for patterns in the images.

Deep Learning: Deep learning is an extension of machine learning that makes use of artificial neural networks to simulate how the human brain learns.

Long Short-Term Memory: Information can last thanks to a deep learning, sequential neural network. By default, LSTM can save the data for a very long time. It is utilized for time-series data processing, forecasting, and classification.

Machine Learning: It is a subfield of artificial intelligence (AI) that focuses on using data and algorithms to mimic human learning processes and progressively increase accuracy.

Parkinson's Disease: This disease affects the neurological system and the areas of the body that are under the control of the nerves. It is a progressive sickness. It results in unintentional or uncontrollable motions including shaking, stiffness, and trouble balancing, and coordinating one's movements.

Random Forest: It is a supervised machine learning system made up of decision trees. Both the classification and regression methods employ Random Forest.

XGBoost: It is a distributed, scalable gradient-boosted decision tree (GBDT) machine learning framework. It offers parallel tree boosting and is the top ma- chine learning package for issues including regression, classification, and ranking.

Chapter 9
Multilevel Vehicle Allotment and Retrieval System Using IoT

Neetha Mudaraddi
Tontadarya College of Engineering, India

Shailaja Mudengudi
https://orcid.org/0000-0001-6102-7942
Tontadarya College of Engineering, India

Muktha Patil
Tontadarya College of Engineering, India

ABSTRACT

This chapter aims to automate the system of vehicle parking with less human interference. This busy world has brought people to a state of no patience, and advancement in technology, fortunately, is favoring the need to save time and energy. The area fitted with sensors provides an efficient way to park the vehicles automatically knowing details of the available slot, which is the target of this project. In addition to this is the IoT implementation to the system which is an advancement in knowing the status of the parking lot even before entering into it. This is illustrated using an application where we get the information about the availability of the slot even before physically entering the parking lot. In addition to it, there will be a multilevel parking facility for the vehicles parked into the lot. These all together comprise the entire working of the project.

DOI: 10.4018/979-8-3693-0786-1.ch009

1. INTRODUCTION

In this 21st century, with all the increasing population, it has become a tedious job to find a parking space for the vehicles. People who go for work every day need a slot near their office parking systems and people who go shopping on weekends need a slot to park at malls, theatres, hotels etc. But in their busy schedule search for a parking place should not become tiresome (Bonde, 2014, pp. 1-4).

In addition, the parking slots are never convenient and provide no logical data about the availability of the slot unless the user visits it manually. These kind of problems are faced regularly by every individual because the factor of unpredictability is very high and there is no existing solution for solving the issue that may benefit the users by saving their time or keeping their mental state happy and carefree (Bonde, 2014, pp. 1-4).

Under all this, increasing traffic results the drivers have to go back and forth desperately looking for parking spaces, wasting their valuable time, fuel consumption, with increased probability of causing accidents. Therefore, to solve all these issues to some extent we propose a multilevel vehicle allotment and retrieval system using IoT. This exciting design of parking system allows the driver to look for available slots and park accordingly. Once the parking is full the gate does not open to any next cars until any spot gets free (Khanna & Anand, 2016).

In this chapter we are going to construct a smart vehicle parking system that connects to internet to help a vehicle driver or any vehicle owner to check if there is a vacant parking spot exist in a parking area even before the driver reach the intended parking slot destination. An IoT based a vehicle parking management system to ease the search for a vacant parking slot in a parking slot through a Smartphone (Khanna and Anand, 2016). The system utilizes several sensors and microcontrollers with internet capability for detecting parked vehicles and to update the data in real-time on internet. The proposed smart parking slot circuit will be equipped with several sensors, inexpensive microcontrollers and Wi-Fi module using which any vehicle owner can check if there is vacant space in a parking slot using his or her phone or tablet or even on computer (Abd Kadir, 2020).

Here we are building an IoT based Vehicle Parking System using NodeMCU, IR sensors, and servo motors. IR sensors are used at the entry and exit gate to locate the vehicle and used to locate the parking slot availability. Servo motors are used to open and close the gates according to the sensor value. The ability to connect, analyze and automate data gathered from devices, powered by and described as the IoT, is what makes smart parking possible (Anwita, 2021).

Smart Parking contains the use of low-cost sensors, real-time data, and applications that allow users to observe available and unavailable parking spots. The goal is to automate and reduce time spent manually searching for the optimal parking floor, spot, and even slot. Whether the destination is a retail store, an airport, or a corporate business office, visitors will surely be impressed with the cutting-edge technology and convenience factors (Seignez & Maurin, 2005).

Overall, Smart Parking is a parking solution that can include in-ground Smart Parking sensors or sensors for counting. These devices are usually embedded into parking spots or positioned next to them to detect whether parking slots are free or occupied. This happens through real-time data collection. The data is then transmitted to a smart parking mobile application or website, which communicates the availability to its users (Seignez & Maurin, 2005).

Nowadays every family is maintaining a car and during the weekends or any festive season every family likes to go shopping malls, Multiplexes, Restaurants, Airports, Railway stations, and Bus stations. Finding a space to park our vehicle had become an annoying issue in those busy places nowadays and who go with their families or friends are wasting more than half an hour to park the vehicle and the vehicles especially the cars are taking up a lot of space. Most of the traffic happens only due to excess vehicles within the metropolitan areas thus people are consuming time in finding areas to park their vehicles (Azshwanth, 2019). And one more issue is also added is pollution while vehicles are in search for slots to park the vehicle and the engine will be in ON state so there will be wastage of fuel.

The Automatic vehicle parking system is designed in order to make the parking process easy and have less man hand. We are implementing IR sensors near the gate to detect the car presence and which intern rotates the servo motor and opens the gate. The display (LCD) shows us the total number of slots and details of their occupancies. Once the car enters and takes a slot, that slot is displayed to be filled. In this fashion the system of vehicle parking is automated (Kamble & Dehankar, 2015).

The idea of the Internet of Things (IoT) began with things with specialized personality gadgets. The gadgets could be followed, controlled, or checked utilizing far off PCs associated through Internet. IoT expands the utilization of Internet, giving the correspondence, and consequently between the system of the gadgets and physical items, or 'Things.'

The Internet empowers the sending, accepting, or conveying of data. Word reference significance of 'Thing' is a term used to refer to a physical item, an activity or thought, circumstance, or action, if when we don't wish to be exact. IoT, in general, comprises of between system of the gadgets and physical objects (Bansal, 2017).

The Internet of things (IoT) is a development of web connectivity in the direction of real gadgets and commonplace phenomena. They are fixed along with electronic components, internet connectivity, furthermore alternative patterns about sensors; the abovementioned gadgets can get through along with reach out alongside excess done with the web furthermore can be controlled and observed remotely. The IoT is interconnecting with real gadgets, automation (referred to as connected devices and wise gadgets), apartments, and alternative items implanted alongside electronics, software, sensors, actuators, and web connectivity whatever is created through the above-mentioned objects to gather and swap information (Satyanarayana &Akhil, 2022).

Everything is particularly encountered through its embedded computing. The system still it can interoperate in a period of the actual web framework. The linkage about these embedded gadgets hold normal towards adoption in automation approximately all ranges, while also permissive leading operation related a wise framework along with spread toward fields in the same manner with smart capital (Vamshi & Tejaswini, 2021). Automatic parking slot booking system for cars has been area of interest of many researchers. People often find difficulties in parking vehicle at public places. Many a times, it happens that, if the slot is already preoccupied, it not only wastes time, but also causes traffic jam at places like theater, play grounds, supermarkets etc. Conventional systems used in India are based on pay and park system. In this system, person has to pay money to park his vehicle at any available slot (Choeychuen, 2013, pp. 117–121).

Traffic congestion in metropolitan areas increases due to the mismanagement of available parking space. To maintain these factors it is important to compose automated smart parking management. This smart parking system is able to help drivers to find parking space easily. Vehicle owners are able to be aware about parking slots. Drivers and motorists are able to generate 30% traffic in metropolitan areas. The Internet of Things is involved in data transmission without human participation. Affordable wireless technology is used by users by IoT. A survey reveals that vehicle population will be increased to 1.6 billion around 2035 and to conduct these vehicles around one million barrels of oil will be burnt. Proper parking space is provided by IoT based smart parking systems that are considered a key solution of wastage of the fuel (Cynthia & Gopinath, 2018).

2. LITERATURE SURVEY

1. "Automated car parking system commanded by Android application''

   Authors: D. J. Bonde, R. S. Shende, K. S. Gaikwad, A. S. Kedari, and A.U. Bhokre.

   Publisher: IEEE Online Website.

   Published On: Proc. Int. Conf. Comput. Commun. Inform. (ICCCI), Volume 5, Issue 2014, pp. 1_4

   Goal: To developaAutomated car parking system using and android.

   Method: Arduino Uno interfaced with LCD and mobile application to command the overall project.

Result: The automated car parking system is modeled.

2. "IoT based smart parking system"
   Authors: A. Khanna and R. Anand.
   Publisher: IEEE Online Website.
   Published On: 2016 International Conference on Internet of Things and Applications (IOTA), January 2016, pp. 266- 270, doi: 10.1109/IOTA.2016.756273.
   Goal: To model a smart parking system using IoT.
   Method: Implemented using Internet of things and applications.
   Result: IoT based smart parking system is modeled.
3. "IoT based Car Parking Management System using IR Sensor"
   Authors: AbdKadir, M. M., Osman, M. N., Othman, N. A., &Sedek, K. A
   Publisher: IEEE Online Website.
   Published On: Journal of Computing Research and Innovation, Volume 05,Issue 02,2020.
   Goal: To model a smart parking system using IR sensors.
   Method: Implemented using IR sensors using IoT as medium of communication.
   Result: IoT based car parking system is modeled.
4. "Smart Parking System Based on the IR Sensor And Node MCU with the Blynk Application"
   Authors: Chakraborty Anwita, Shah Chirag Shrey, Kumar SravanaYelithoti, SamantTapaswini, Swayamsiddha Swati.
   Publisher: IEEE Online Website.
   Published On: International Journal of Sensors, Wireless Communications and Control 2021,Volume:11,Issue:4.
   Goal: To propose a smart parking system based on IR sensor and NodeMCU with the blynk application.
   Method: Implemented using IR sensors and NodeMCU using blynk application.
   Result: Smart Parking System is modelled based on IR sensors and NodeMCU with blynk application.
5. "Autonomous parking carrier for intelligent vehicle"
   Authors: E. Seignez, A. Lambert and T. Maurin.
   Publisher: IEEE Online Website.
   Published On: Intelligent Vehicles Symposium, 2005., 2005, pp. 411-416, doi: 10.1109/IVS.2005.1505138.
   Goal: To model an autonomous parking carrier for intelligent vehicle.
   Method: Implemented using IoT and IR sensors in association with NodeMCU.
   Result: Automated parking carrier is successfully modeled.
6. "Automated car parking system"
   Authors: D. Azshwanth, MithulTitten Koshy and Mr.T. Balachander.

Publisher: IEEE Online Website.

Published On: International Conference on Physics and Photonics Processes in Nano Sciences 2019, Volume 1362.

Goal: To model an automated car parking system using Arduino UNO and NodeMCU.

Method: Arduino Uno interfaced with LCD and mobile application to command the overall project.

Result: The automated car parking system is successfully modeled.

3. PROBLEM STATEMENT

The increasing number of vehicles in metropolitan areas has led to a significant challenge of finding parking space. This often results in congestion, frustration, and waste of time and fuel. To overcome this challenge, an IoT-based smart car parking system can be implemented to provide drivers with real-time information about availability of parking slots, thus reducing the time and effort needed to find a parking space.

Objectives

The objectives for an IoT-based parking system can vary depending on the specific needs and goals of the organization or community implementing it.

- **Optimizing Parking Space Utilization:** One of the main objectives of an IoT-based parking system is to optimize the utilization of parking spaces. By monitoring parking spots in real-time, the system can provide precise information about available parking spaces, reducing the time spent searching for a parking spot.
- **Improved Traffic Flow:** An IoT-based parking system also helps improve traffic flow around parking areas. By providing information about available parking spaces, the system can reduce the number of cars roaming around looking for a parking spot, reducing traffic congestion and emissions.
- **Increased Safety:** IoT-based parking systems can also increase safety by providing real-time information about the occupancy of parking spaces. Drivers can avoid accidents by knowing when a parking spot is occupied or empty.
- **Reduced Environmental Impact:** By optimizing parking utilization, reducing traffic and emissions. IoT-based parking systems can contribute to reducing the environmental impact of transportation.

- **Enhanced User Experience:** An IoT-based car parking system can enhance the user experience by providing convenient and efficient parking solutions. With real-time parking availability information, drivers can find easily the parking spots and make better use of their time.
- **Efficient Revenue Collection:** IoT-based parking systems can also provide efficient revenue collection by automating the payment process. With an integrated payment system, drivers can easily pay for the parking using mobile apps, reducing the need for cash transactions and manual processes.
- **Remote Monitoring and Management:** An IoT-based parking system can also provide remote monitoring and management capabilities. The system can be monitored and managed remotely, providing real-time information about parking occupancy, revenue collection, and other critical metrics. This can help operators make data-driven decisions and optimize parking operations.
- **Sustainability:** The system is designed with sustainability in mind, using renewable energy sources, minimizing waste, and reducing the carbon footprint.
- **Data analytics:** The system should collect and analyze data on parking patterns, usage, and trends, providing insights that can be used to optimize the system and improve the user experience.
- **Seamless integration:** The system should be seamlessly integrated with other smart city infrastructure, such as public transportation systems and traffic management systems, to create a cohesive and efficient urban environment.

4. ORGANIZATION OF CHAPTER:

The chapter will be divided into several stages, including:

- **Design and development of the IoT-based smart parking system:** This stage will involve the development of the hardware and software components required for the system, including sensors, communication protocols, and data analytics tools.
- **Testing and validation of the system:** The system will be tested in a controlled environment to ensure its accuracy and reliability. The testing will involve simulating various parking scenarios and validating the system's ability to detect and report available parking spaces.
- **Deployment and integration of the system:** The system will be deployed in a real-world environment and integrated with existing parking infrastructure. This will involve the installation of sensors in parking spots, the configuration of communication protocols, and the integration of data analytics tools.

- **Data analysis and optimization:** The data collected from the system will be analyzed to identify trends and patterns in parking usage. This information can be used to enhance parking management, improve resource allocation, and reduce the number of parking violations.

Overall, this chapter aims to develop an innovative and practical solution to the problem of parking management in metropolitan areas. By leveraging IoT technology and data analytics, the system will provide real-time information to drivers and parking authorities, enabling more efficient and effective use of parking resources.

5. PROPOSED METHODOLOGY

Block Diagram

The above block diagram is the depiction of how automated vehicle parking system works. We have used IR sensors for the detection of vehicles and Adafruit IO server for displaying the status of the parking slot. Further servo motor is used as gate or barrier for entering or exit of vehicles and the operation is controlled by ESP8266 where the power supply is the driving force of the project.

These infrared sensors are connected with the ESP8266. So when a vehicle is detected in the slot, it sends a command to the Nodemcu ESP8266 Wi-Fi module, then Nodemcu sends the command to the Adafruit IO server. In the sever it is available both in web page and mobile application. The details are displayed over the server as the hardware and software is integrated with each other.

Where parking slot will be filled when the vehicle are parked in the slot and if the remaining slots are empty then the status of the parking slot will be available in the Adafruit IO server so that the other user can park the vehicle in the available parking slot. The overall chapter is showcased in the block diagram, at every stage the data that is the information of the availability of the slots for parking the vehicle

Figure 1. Project block diagram

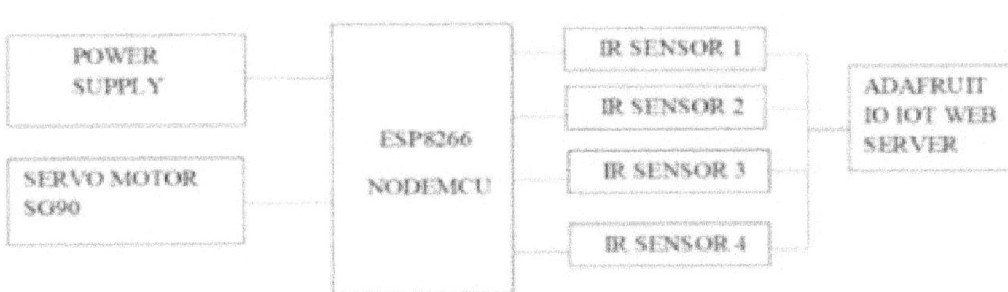

is important and this data is real time.The collected data is communicated through the IoT medium which is the soul purpose of the chapter and online platform so that the entire process is automated made easier for user with less man handle.

Flow Chart

The above flow diagram shows the entire working of our project in a step-to-step process. Initially when any vehicle arrives at the gate where we have fixed the sensors to detect them, the gate opens upon the availability of slots in the parking lot and otherwise remains closed. The IoT implementation of the information of slots available directly in an application which the user can access from anywhere so that it saves the time and avoids the user from reaching the parking lot if there is no availability of slots.

Another advantage is that vehicles can be parked in the lot at multiple stages, that is we are implementing upliftment of vehicles for the availability of slots. The upliftment facility refers to parking the vehicles at multiple stages in the parking lot to provide more space for the vehicles to be parked. Every time the slots are filled in the parking lot the gate does not open and the vehicles have to be moved to other available parking lots nearby. But with ease of IoT implementation we can access

Figure 2. Working flow of the project

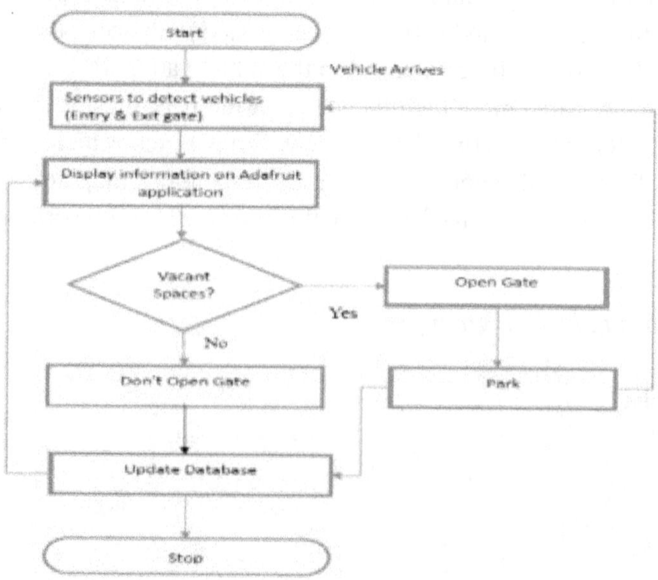

the information about the availability of slots even before reaching the parking lot which is main target of this project.

The proposed methodology for an IoT-based smart parking system involves the following steps:

- **Design and development of the hardware components:** This step involves the design and development of the hardware components required for the system, including sensors, microcontrollers, communication modules, and power supplies. The sensors will be placed in each parking spot to detect the presence of a vehicle and transmit this information to the microcontroller.
- **Installation of the hardware components:** The hardware components will be installed in each parking spot, and the communication module will be connected to a local network. The local network will be connected to the internet, allowing the data to be transmitted to the central server.
- **Integration with existing parking infrastructure:** The IoT-based smart parking system will be integrated with the existing parking infrastructure, including the parking meters, payment systems, and signage. This integration will allow the system to provide real-time information about available parking spaces, pricing, and payment options.
- **Data analysis and visualization:** The data collected from the sensors will be processed and analyzed using data analytics tools, such as machine learning algorithms, to identify patterns and trends in parking usage. The results will be visualized using dashboards and reports, providing valuable insights for parking management.
- **Implementation of smart parking features:** The system can be further enhanced by implementing smart parking features, such as reservation systems, parking guidance, and automatic payment systems. These features can be integrated with the existing infrastructure to provide a seamless parking experience for drivers.

Overall, the proposed methodology for an IoT-based smart parking system involves the design and development of hardware and software components, installation of the hardware components, integration with existing parking infrastructure, data analysis and visualization, and implementation of smart parking features. The system can provide real-time information about parking availability, improve parking management, and enhance the parking experience for drivers.

6. REQUIREMENTS

Hardware Components:

1. ESP8266-NodeMCU

ESP8266 is a budget friendly Wi-Fi microchip with full TCP/IP stack and microcontroller capability. It was developed by the Chinese manufacturer Espressif Systems and released in 2014. The chip features a 32-bit RISC CPU running at 80 MHz; 64KB of instruction RAM and 96KB of data RAM, and is capable of connecting to Wi-Fi networks and interacting with other devices over the internet. The ESP8266 has become in demand for its ease of use and low cost, making it an ideal choice for hobbyists and professionals alike. It can be programmed using the Arduino IDE and supports a wide range of sensors and actuators, making it an ideal choice for IoT (Internet of Things) projects. Additionally, the ESP8266 can be used as a standalone microcontroller or as a WiFi module in conjunction with other microcontrollers such as Arduino or Raspberry Pi.

Figure 3. NodeMCU-ESP8266 (WifiModule)

2. Infrared Sensors (IR)

An IR (infrared) sensor, also known as IR receiver, is a device that detects and responds to infrared radiation. Infrared radiation is a type of electromagnetic radiation that is not visible to the naked eye, but can be detected and measured by an IR sensor. IR sensors are commonly used in a various applications, such as remote controls, motion detectors, and temperature sensors. In a remote control, for example, the IR sensor detects the infrared radiation emitted by the remote control and translates it into electrical signals which are sent to the device being controlled. In a motion detector, the IR sensor detects changes in the amount of infrared radiation in its field of view and triggers an alarm or other response.

Figure 4. Infrared sensor

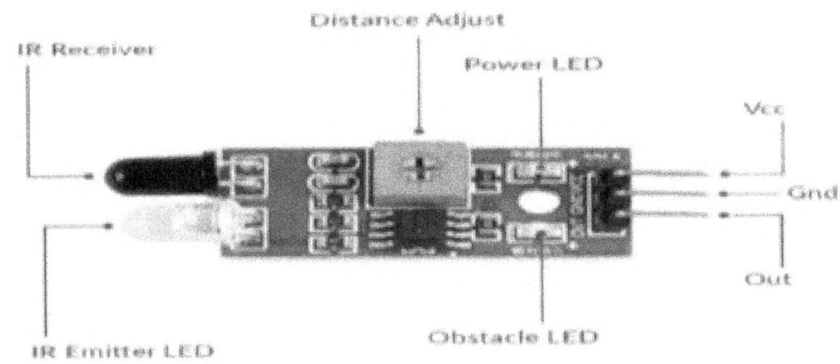

An IR (infrared) sensor, also known as IR receiver, is a device that detects and responds to infrared radiation. Infrared radiation is a type of electromagnetic radiation that is not visible to the naked eye, but can be detected and measured by an IR sensor.IR sensors are commonly used in a variety of applications, such as remote controls, motion detectors, and temperature sensors. In a remote control, for example, the IR sensor detects the infrared radiation emitted by the remote control and translates it into electrical signals that are sent to the device being controlled. In a motion detector, the IR sensor detects changes in the amount of infrared radiation in its field of view and triggers an alarm or other response.

3. Servomotor SG 90

The SG90 servo motor is a small, lightweight servo motor commonly used in hobby projects and small-scale robotics. It is designed to rotate between 0 and 180 degrees, and is controlled using pulse width modulation (PWM) signals from a microcontroller or other control device. The SG90 servo motor has a maximum torque of 1.8 kg/cm and operates on a voltage range of 4.8V to 6V DC. It typically draws less than 250mA of current when under load, making it a good choice for battery-powered applications.

The SG90 servo motor is popular in hobbyist projects due to its low cost, ease of use, and small size. It is used for applications such as remote control cars, small robots, and model airplanes. However, due to its relatively low torque output and plastic gears, it may not be suitable for larger or more demanding applications.

Figure 5. Servomotor SG 90

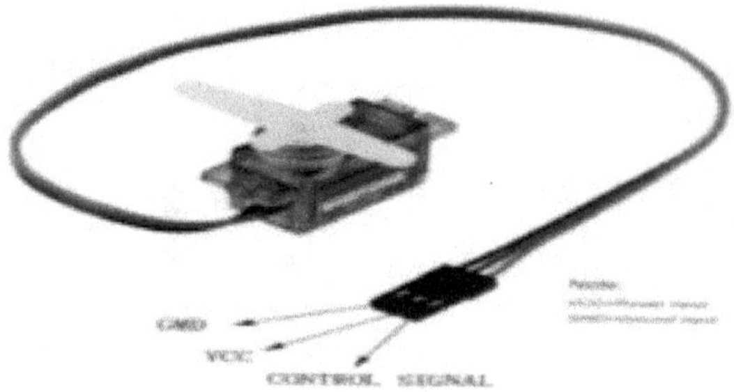

4. Male and Female Wires

A male-male wire has a male connector on each end. A male connector is typically a pin or a plug with one or more exposed metal contacts that are designed to fit into a corresponding female connector. Male-male wires are often used to connect two components that both have female connectors.

A female-female wire has a female connector on each end. A female connector typically has one or more recessed openings that are designed to accept male connectors. Female-female wires are often used to extend the length of a male-male wire or to connect two components that both have male connectors.

A male-female wire has a male connector on one end and a female connector on the other. Male-female wires are often used to connect a component with a male connector to a component with a female connector.

Figure 6. Male-male, male-female, and female-female wires

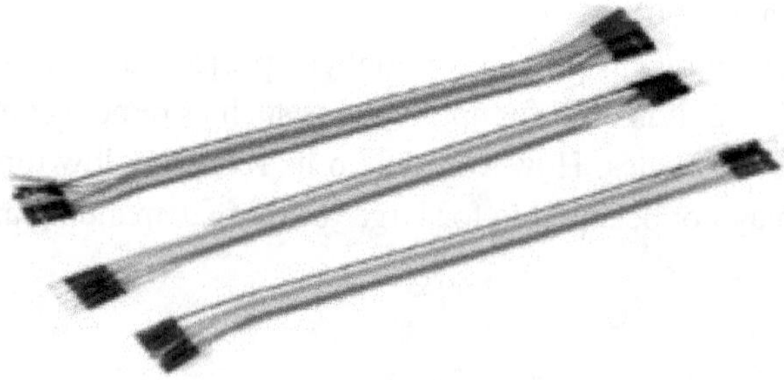

Figure 7. PCB board

5. Printed Circuit Board

A Printed Circuit Board (PCB) is a board made of insulating material, such as fiberglass or plastic that is used to connect electronic components together. The board has thin metal strips, usually made of copper, that create a circuit path between the components. The circuit is printed onto the board using a special etching process. The board is coated with a layer of copper, and then a protective layer of photoresist is applied.

The photoresist is exposed to light through a mask, which creates a pattern on the board. The unexposed photoresist is then washed away, leaving behind the pattern of the circuit. The board is then etched in a chemical bath, which removes the copper from the areas not covered by the photoresist, leaving the circuitry behind. PCBs are used in a wide variety of electronic devices, from simple toys and gadgets to complex computer systems and medical equipment. They offer a reliable and efficient way to connect components and create complex circuits in a compact space.

6. Power Supply

A 5V 2A power supply is a type of electrical power supply which provides a constant voltage of 5 volts and a maximum output current of 2 amps. This type of power supply is commonly used to power small electronic devices that require a low voltage and a moderate amount of current. Examples of devices that may use a 5V 2A power supply include smartphones, tablets, and some small single-board computers like the Raspberry Pi.

It is important to use the correct power supply for your device, as using a power supply with a different voltage or current output could damage your device or cause it to malfunction. When selecting a power supply, it is important to check the voltage

Figure 8. Example of Embedded C program

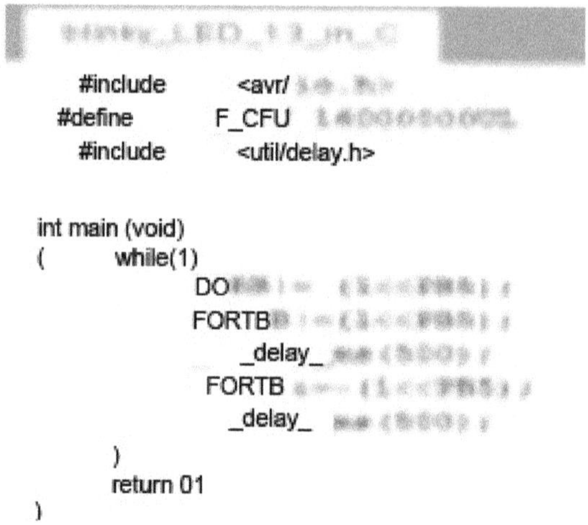

and current requirements of your device to ensure that the power supply can provide the necessary power.

Software Components

Embedded C-Programming Language

Embedded C is a programming language that is specifically designed for programming microcontrollers and other embedded systems. It is a variant of the C programming language, with features and optimizations that are specific to the needs of embedded systems. Embedded C is used to write low-level code that interacts with hardware and peripherals, such as sensors, motors, and displays.

The language provides low-level access to the microcontroller's hardware, such as memory-mapped I/O and control registers, allowing for precise control of the system's behavior. Embedded C code is often written in a text editor and compiled using a cross-compiler, which generates machine code for the specific microcontroller or embedded system.

8. Arduino IDE

The Arduino Integrated Development Environment (IDE) is a software application that is used to develop software for Arduino microcontroller boards. It provides a complete development environment for writing, compiling, and uploading code to an Arduino board. The Arduino IDE is available for Windows, Mac OS X, and Linux.

Figure 9. Arduino IDE interface

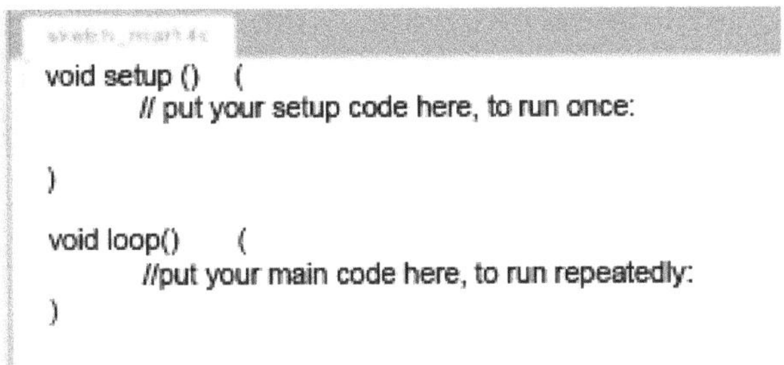

It includes a text editor with features such as syntax highlighting, code completion, and error highlighting.

The IDE also includes a compiler and a bootloader, which allows the code to be uploaded to the Arduino board through a USB cable. The IDE is designed to be easy to use, even for those without a programming background. It includes a library of example code and tutorials to help users get started. The Arduino IDE also supports a variety of third-party libraries, which can be easily installed and used to add additional functionality to the Arduino board.

9. Adafruit IO

Adafruit IO is an Internet of Things (IoT) platform developed by Adafruit Industries, a company that produces open-source hardware and software for makers and

Figure 10. Overview of Adafruit IO

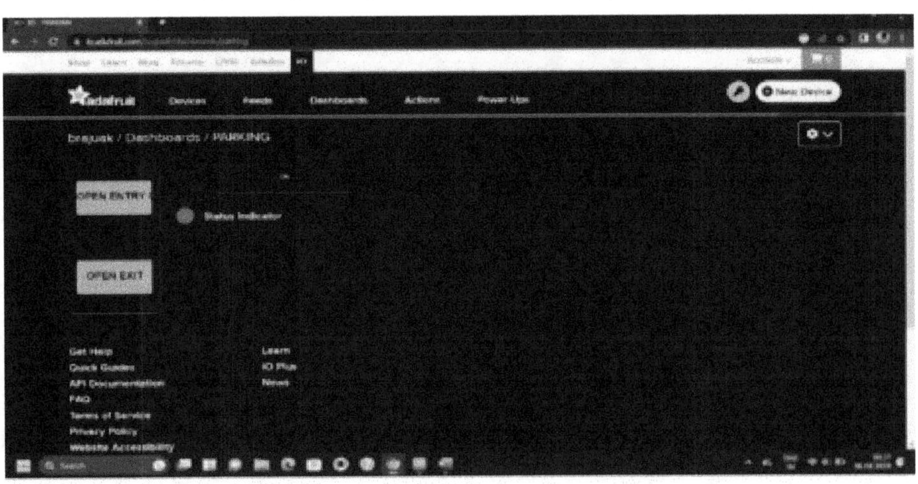

hobbyists. Adafruit IO is a cloud-based platform that allows users to easily connect and interact with their IoT devices and data. With AdafruitIO users can create and manage data feeds, which are streams of data from their IoT devices. These feeds can be monitored and controlled through the Adafruit IO dashboard, a web-based user interface that allows users to visualize their data, set up triggers and alerts, and control their devices remotely.

7. ADVANTAGES, DISADVANTAGES, AND APPLICATIONS

Advantages

1. **Increased efficiency:** Smart parking systems use sensors and IoT technology to provide real-time information about parking space availability. This can reduce the time drivers spend searching for a parking spot and improve overall efficiency.
2. **Improved user experience:** Smart parking systems provide drivers with real-time information about parking availability, location, and pricing, which can improve the overall experience and reduce frustration.
3. **Increased revenue:** Smartcar parking systems can generate revenue by charging for parking, reducing the number of unpaid tickets and fines, and improving overall parking management.
4. **Reduced traffic congestion:** Smart parking systems can reduce traffic congestion by directing drivers to available parking spaces and reducing the number of vehicles searching for parking.
5. **Enhanced safety and security:** Smart parking systems can improve safety and security by monitoring parking spaces and detecting unauthorized access or suspicious activity.

Disadvantages

1. **Cost**: Implementing an IoT-based smart parking system can be expensive, as it involves the installation of sensors, cameras, and other hardware. This can make it challenging for some organizations to justify the investment.
2. **Maintenance:** Smart parking systems require regular maintenance and upgrades to ensure that they continue to function properly. This can be time-consuming and costly, particularly for organizations with limited resources.
3. **Technical expertise:** Implementing and maintaining an IoT-based smart parking system requires technical expertise, which may not be available in all

organizations. This can make it challenging to install and operate the system effectively.

4. **Data privacy concerns:** Smart parking systems collect and store data about vehicles and drivers, which can raise privacy concerns if the data is not properly secured. This is particularly true if the data is shared with third parties or used for purposes other than parking management.
5. **Reliance on technology:** IoT-based smart parking systems rely on technology, which can be subject to failure or disruption. This can lead to problems with the system's operation and reduce its effectiveness.

Applications

1. **Urban areas:** Smart parking systems are commonly used in urban areas with high traffic volumes and limited parking space. They can help reduce congestion, improve parking efficiency, and enhance the overall parking experience.
2. **Commercial and retail areas:** Smart parking systems can be used in commercial and retail areas to improve customer experience and generate revenue. They can provide real-time information on parking availability, location, and pricing, making it easier for customers to find a parking spot.
3. **Public transportation:** Smart parking systems can be used to improve access to public transportation by directing drivers to available parking spaces near public transportation hubs. This can help reduce traffic congestion and encourage more sustainable transportation options.
4. **Hospitals:** Smart parking systems can be used in hospitals to improve patient access and reduce congestion in parking lots. They can provide real-time information on parking availability and location, making it easier for patients and visitors to find parking spots.
5. **Airports:** Smart parking systems can be used in airports to improve the overall parking experience for travelers. They can provide real-time information on parking availability, location, and pricing, making it easier for travelers to find a parking spot and reducing the time they spend searching for a spot.

8. IMPLEMENTATION

Steps involved in implementing our chapter are as follows:

- **Planning:** The first step in implementing a smart parking system is to develop a plan for the system, including its scope, objectives, and requirements. This may involve conducting a site survey to identify potential parking locations and assess existing parking infrastructure.

- **Hardware installation:** Once the plan is developed, the next step is to install the necessary hardware, including sensors, cameras, and other IoT devices. This may require wiring and other infrastructure modifications to connect the devices to a central system.
- **Software development:** Once the hardware is installed, the next step is to develop the necessary software to collect and process data from the IoT devices. This may involve developing custom software or using existing parking management software that is compatible with the IoT devices.
- **Integration:** The software and hardware components of the smart parking system must be integrated to ensure that the system functions as intended. This may involve developing custom integration software or using existing integration tools.
- **Testing:** Before the system can be deployed, it must be thoroughly tested to ensure that it is functioning properly and meets all requirements. This may involve a combination of unit testing, integration testing, and user acceptance testing.
- **Deployment:** Once the system is tested and approved, it can be deployed in the target environment. This may involve training users and stakeholders on the system and providing ongoing support and maintenance.

9. OUTCOME OF PROJECT

The following images are the results or outcome of our project, the images depict the overview of our entire chapter.

Figure 11. The chapter model

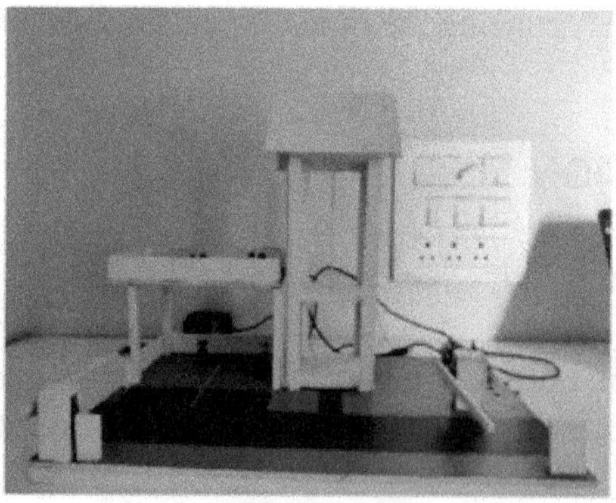

Figure 12. Vehicle entry point

Figure 13. Vehicle exit point

Figure 14. The multilevel parking facility

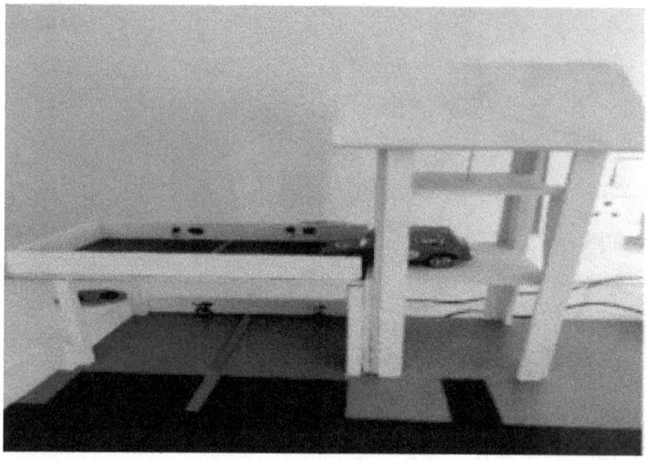

Figure 15. The initial status of the system

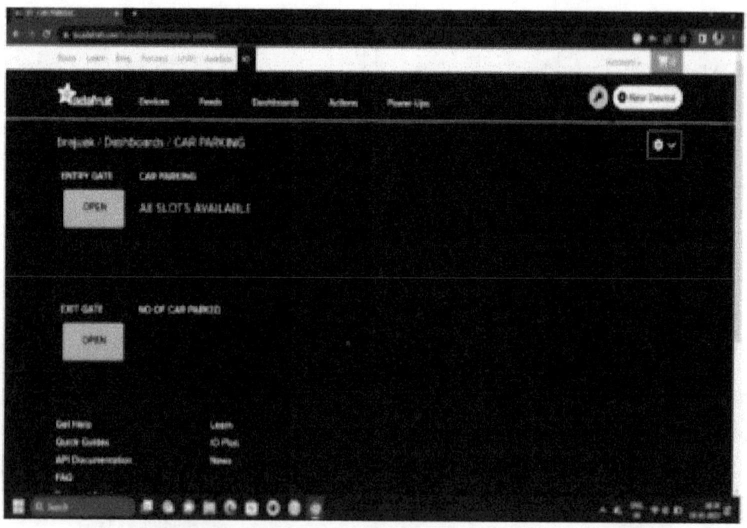

Figure 16. The updated status 1

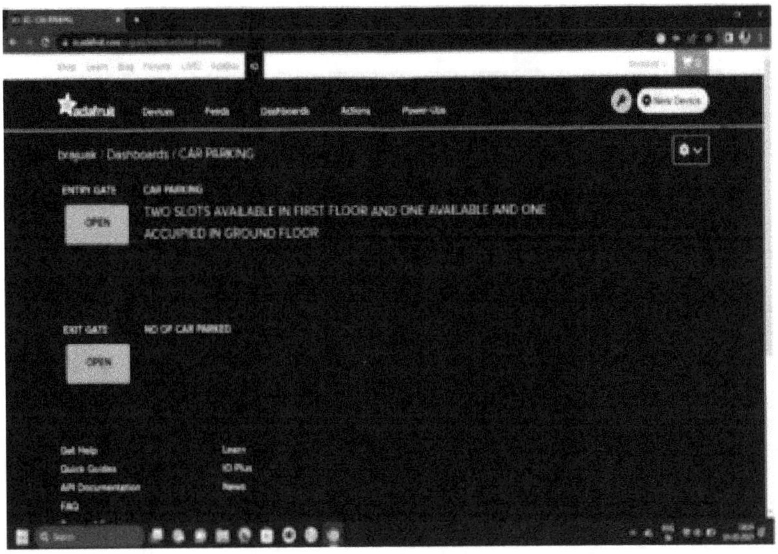

Figure 17. The updated status 2

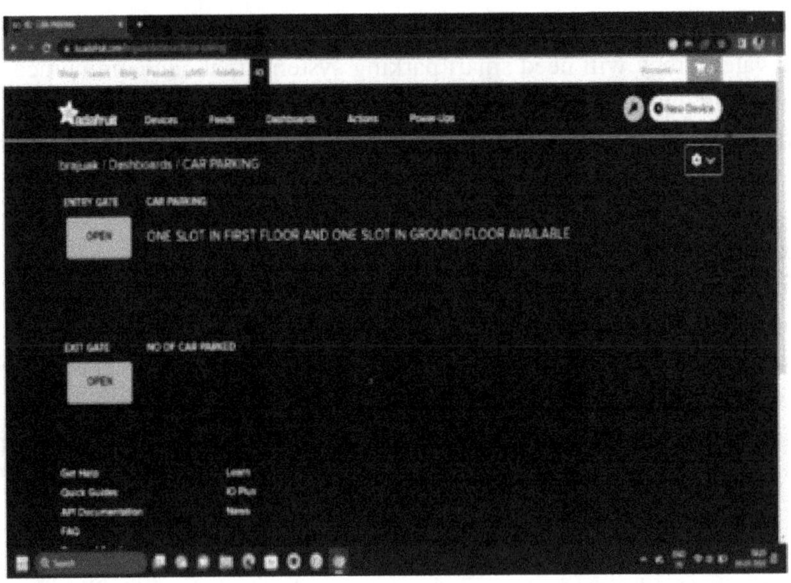

Figure 18. The updated status 3

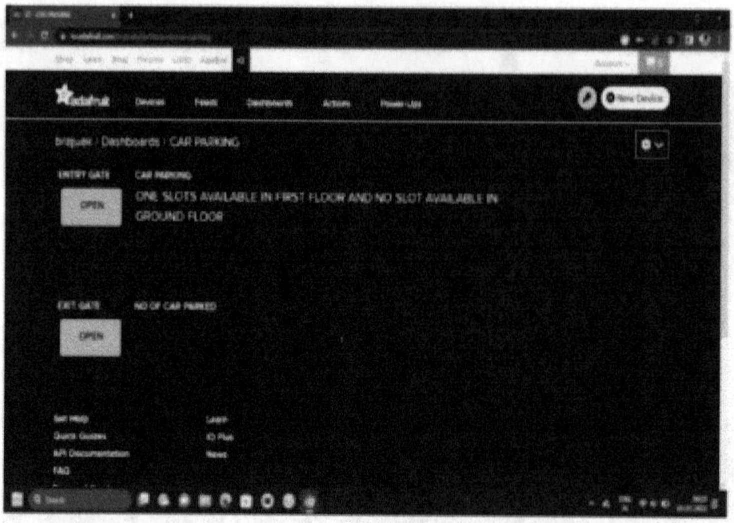

FUTURE SCOPE

- **Integration with autonomous vehicles:** As self-driving cars become more prevalent, they will need smart parking systems that can communicate with them and guide them to available parking spots.
- **Real-time data analytics:** Smart parking systems can collect and analyze real-time data on parking usage patterns, occupancy rates, and vehicle flows to optimize parking operations, reduce congestion, and improve customer experience.
- **Seamless payment and billing:** IoT-based smart parking systems can integrate with payment gateways and mobile wallets, making it easier for users to pay for parking and enabling operators to streamline billing and revenue management
- **Sustainability and energy efficiency:** Smart parking systems can incorporate renewable energy sources, such as solar panels, and use energy-efficient lighting and sensors to reduce their environmental impact.
- **Improved security and safety:** IoT-based smart parking systems can incorporate advanced security features, such as video surveillance, access control systems, and emergency response mechanisms, to enhance the safety and security of parked vehicles and their occupants.

- Sustainability and energy efficiency: Smart parking systems can incorporate renewable energy sources, such as solar panels, and use energy-efficient lighting and sensors to reduce their environmental impact.
- Improved security and safety: IoT-based smart parking systems can incorporate advanced security features, such as video surveillance, access control systems, and emergency response mechanisms, to enhance the safety and security of parked vehicles and their occupants.

CONCLUSION

The Multilevel vehicle allotment and retrieval parking system automation is successfully modeled. The mind numbing process of today is made simple with our project design. The working of our project is at the stage of vehicle allotment and retrieval based upon the availability of slots in the parking slot through Adafruit IO server. The visitors near the parking slot will be allowed inside through the servo barrier upon detecting the vehicles in conjunction with availability of slots in the parking slot. An upliftment provision is made for the same to provide more space so that the vehicles can be parked easily in comfortable vacant spaces. The IoT implementation is the heart of our project as the user need not visit the parking slot physically and can look for the vacant space before reaching the parking slot.

REFERENCES

AbdKadir. (2020). IoT based Car Parking Management System using IR Sensor. *Journal of Computing Research and Innovation*, *5*(2).

Anwita, Shrey, Yelithoti, Tapaswini, & Swati. (2021). Smart Parking System Based on the IR Sensor And Node MCU with the Blynk Application. *International Journal of Sensors, Wireless Communications and Control, 11*(4).

Azshwanth, D. (2019). Automated car parking system. *International Conference on Physics and Photonics Processes in Nano Sciences,* 1362.

Bonde, D. J., Shende, R. S., Gaikwad, K. S., Kedari, A. S., & Bhokre, A. U. (2014). Automated car parking system commanded by Android application. *Proc. Int. Conf. Comput. Commun. Inform. (ICCCI),* 5, 1-4.

Chaudhary, H., Bansal, P., & Valarmathi, B. (2017). Advanced Car parking system using Arduino. *4th International Conference on Advanced Computing and Communication System*, 1-5.

Choeychuen, K. (2013). Automatic parking lot mapping for available parking space detection. *Proceedings of the 5th International Conference on Knowledge and Smart Technology (KST),* 117–121.

Cynthia, Priya, & Gopinath. (2018). IOT based Smart Parking Management System. *International Journal of Recent Technology and Engineering, 7*(4S), 374-379.

Kamble & Dehankar. (2015). Review on Automatic Car Parking Indicator System. *Journal on Recent and Innovation Trends.*

Khanna, A., & Anand, R. (2016). IoT based smart parking system. *International Conference on Internet of Things and Applications (IOTA),* 266-270. doi: 10.1109/IOTA.2016.756273

Laugier, Fraichard, Garnier, & Paromtchik. (1999). *Sensor-based control architecture for a car-like vehicle.* Springer.

Multilevel Vehicle Allotment And Retrieval System Using IoT. (2022). *International Research Journal of Modernization in Engineering Technology and Science*, 4(3).

Noor, N. M., Razak, Z., & Yamani, M. (2009). Car Parking System: A Review of Smart Parking System and its Technology|| -. *Information Technology Journal.*

Satyanarayana, S. K., Akhil, A., & Padmini, G. (2021). Smart parking assistance using Arduino. *International Journal for Research in Applied Science & Engineering Technology, 9*(6).

Seignez, E., Lambert, A., & Maurin, T. (2005). Autonomous parking carrier for intelligent vehicle. *IEEE Proceedings. Intelligent Vehicles Symposium*, 411-416. doi: 10.1109/IVS.2005.1505138

Suvarna, S. (2017). Smart Car Parking System using Arduino UNO. *International Journal of Computer Applications, 169*(1), 13-18. DOI: doi:10.5120/ijca2017914425

Chapter 10
Routing in Wireless Sensor Networks and Internet of Things:
Systematic Analysis and Discussion

Yogesh Patidar
https://orcid.org/0000-0002-4803-0968
Mandsaur University, India

Manish Jain
Mandsaur University, India

Ajay Kumar Vyas
https://orcid.org/0000-0002-2462-2838
Adani University, India

ABSTRACT

In this chapter, the systematic analysis and literature of the existing routing protocols are discussed. The application-specific choice of the routing method is a primary challenge in a wireless sensor network or internet of things for efficient and reliable transmission of data. The reliable transmission of the data depends on various factors such as application, data generation rate, sampling rate, etc. The most commonly used routing is flat routing, also called data-centric routing. There are many other routing methods, e.g., hierarchical routing, location-based routing, energy-aware routing, multipath routing, etc. In this chapter, the discussion is divided into purely proactive and reactive routing methods or clustering-cum-routing advanced methods for all the different applications and needs. In the chapter, the network performance is optimized for the number of optimal clusters, the maximal throughput, and the lifetime. In addition, the comparative analysis of different routing and clustering protocols is also discussed in the chapter.

DOI: 10.4018/979-8-3693-0786-1.ch010

1. INTRODUCTION

Routing and clustering protocols are essential in improving the efficient transmission of data in wireless sensor networks (WSN) and the Internet of Things (IoT). This chapter discusses the energy-efficient protocols and methods used to reduce energy consumption in WSNs, including harvesting, data aggregation, duty cycling, and routing (Fradj, et al., 2018; Javidi & Buhler, 2016; Sun, et al., 2010; Zang, et al., 2010; J, Ajmani, et al., 2021; Wang, 2021; Farooq & Kunz, 2017; Shabbir & Hassan, 2017; Khoshimov, 2019; Tsai, et al., 2020; Mabrouki, et al., 2021). The proposed work also explores the essential applications of energy-efficient WSNs in various fields, such as agriculture field monitoring, healthcare, and smart cities (Xu, et al., 2019; Behera, et al., 2019). Wireless Sensor Networks (WSNs) are an essential component of the Internet of Things (IoT) ecosystem, which enables the collection and transmission of data from various sensor units. However, the limited energy resources of WSN nodes pose a significant challenge to developing sustainable and long-lasting networks (Clare, et al., 1999). In this context, energy-efficient methods play a crucial role in optimizing the performance and lifetime of WSNs. The researchers have reported various protocols for energy-efficient data transmission. The energy can be saved by various methods (Mody, et al., 2021; Heinzelman, et al., 2000; Leu, et al., 2015; Younis & Fahmy, 2004; Panchal, et al., 2020; Ding, et al., 2005; Saranya, et al., 2018; Zhixiang & Bensheng, 2007; Handy, et al., 2002; Lotf, et al., 2008; Sharawi & Emary, 2017; Patidar, et al., 2022b; Hoang, et al., 2010; Zagrouba & Kardi, 2021; M, et al., 2020; Agbehadji, et al., 2019; Yousif et al., 2017). The researchers have proposed energy-efficient medium access control (MAC) protocols, routing methods (Abbasi & Younis, 2007; Kumar, et al., 2019), and energy harvesting methods for energy-efficient data transmission. The remainder of the book chapter is divided into multiple sections. Section 2 discusses the literature study. Section 3 discusses a systematic analysis of various routing methods, and Section 4 describes the contribution of various authors.

2. LITERATURE SURVEY

Title & published at	"*Energy Consumption for Opportunistic Routing Algorithms in WSN*," 2018 IEEE 32nd International Conference on Advanced Information Networking and Applications
Authors	H. Ben Fradj, R. Anane and R. Bouallegue
Objective	This study aims to introduce a unique opportunistic routing method, ECS-OR (Energy Candidate Set-opportunistic Routing) (Fradj, et al., 2018), to address the energy consumption and network longevity concerns in wireless sensor networks (WSNs).
Contribution	This study introduces ECS-OR, a ground-breaking opportunistic routing technique created to maximize energy efficiency and increase the longevity of wireless sensor networks. By offering a cutting-edge method for balancing energy usage in WSNs with constrained non-rechargeable battery power, it makes a contribution to the field. (Fradj, et al., 2018).

Findings	Numerous simulations show that ECS-OR outperforms current WSN routing algorithms, greatly enhancing wireless connectivity and energy efficiency (Fradj, Anane, & Bouallegue, 2018).The results highlight the value of ECS-OR in improving WSN performance generally, especially in situations when sensor nodes have limited resources (Fradj, et al., 2018).
Conclusion	In conclusion, ECS-OR is a viable approach to resolving issues with energy use and network longevity in wireless sensor networks. This research makes a significant contribution by developing a useful tool for increasing the energy efficiency and dependability of WSNs, which has significant advantages for battery-constrained sensor network applications (Fradj, et al., 2018).
Title & published at	*Opportunistic Routing in Wireless Networks,* Publisher: Now Foundations and Trends
Authors	Tara Javidi; Eric Van Buhler
Objective	To highlight its benefits over traditional fixed-path routing techniques, this study intends to investigate the importance of opportunistic routing in wireless multi-hop networks. It also acts as a through resource for academics and researchers interested in opportunistic routing's technical underpinnings (Javidi & Buhler, 2016).
Contribution	The evolution of wireless multi-hop networks from military to other industrial use cases is highlighted in the article. It highlights the shortcomings of fixed-path routing and the demand for opportunistic routing that takes advantage of wireless communication's broadcast nature (Javidi & Buhler, 2016).The research also provides several opportunistic routing algorithms that regularly outperform conventional shortest-path routing techniques in a variety of contexts (Javidi & Buhler, 2016).
Findings	The results emphasize that in wireless multi-hop networks, opportunistic routing techniques outperform fixed-path routing. These methods lessen pointless packet retransmissions, which improves network performance by making online relay decisions based on actual transmission results and node rankings (Javidi & Buhler, 2016).
Conclusion	The adoption of opportunistic routing is encouraged by this study, which also provides researchers and students with a helpful resource for learning more about the technical nuances of the technique.
Title & published at	*"A load-balanced and energy-aware routing metric for wireless multimedia sensor networks,"* IET 3rd International Conference on Wireless, Mobile and Multimedia Networks (ICWMNN 2010)
Authors	W. Sun, Y. Song and M. Chen
Objective	In the context of routing protocols, this paper discusses the difficulties presented by wireless multimedia sensor networks (WMSNs). It focuses particularly on WMSNs that are based on IEEE 802.11s and incorporate MAC-layer mesh networking protocols to offer high coverage, self-organization, and reliability (Sun, et al., 2010).
Contribution	Based on the IEEE 802.11s draft standard, the study offers a novel routing metric called load balanced airtime with energy awareness (LBA-EA) (Sun, Song, & Chen, 2010). In IEEE 802.11s-based WMSNs, LBA-EA evaluates link quality as well as load balancing and energy usage. This ground-breaking statistic makes a substantial contribution to raising these networks' performance (Sun, et al., 2010).
Findings	The efficiency of LBA-EA in extending the lifespan of WMSNs and improving average network throughput is shown by simulation results. Critical issues with WMSNs are addressed by optimizing the network's resource utilization and energy efficiency through the incorporation of load balancing and energy awareness into the routing metric (Sun, et al., 2010).
Conclusion	A unique feature of the proposed routing metric, LBA-EA, is its capacity to enhance network performance by taking link quality, load balancing, and energy consumption into account (Sun, et al., 2010).
Title & published at	*"A Robust Routing Protocol in wireless sensor network,"* IET International Conference on Wireless Sensor Network 2010
Authors	Zhe Zang, J. -D. Qi and Yong-Jie Cao,
Objective	With a primary focus on achieving dependability, low operating costs, and ease of maintenance, this study focuses on the crucial research field of creating routing protocols for wireless sensor networks (WSNs). By presenting a brand-new hop-based routing protocol, HBRRP (Hop Based Robust Routing Protocol for WSN), and offering a thorough examination of hop-based routing protocol concepts, it seeks to satisfy these design goals (Zang, et al., 2010).
Contribution	By suggesting HBRRP, a new routing strategy for WSNs that makes use of hop-based routing principles, this work makes a contribution to the area (Zang, et al., 2010). HBRRP prioritizes parents and siblings as forwarding nodes during data transmission, uses a formula for evaluating the routing quality, and integrates a trigger update mechanism to maintain a dynamic network architecture while preventing "routing holes." Additionally, it lowers the cost of network maintenance by introducing data packets as a component of the routing process (Zang, et al., 2010).
Findings	HBRRP outperforms existing protocols like MHC, DD, and join-MHC in simulations utilizing the OMNeT++ simulation tool when it comes to load balancing, extending network lifetime, dependability, and lowering routing overhead. These results demonstrate HBRRP's effectiveness in accomplishing the stated design objectives for WSN routing protocols (Zang, et al., 2010).

Conclusion	In summary, this study introduces the cutting-edge hop-based routing protocol HBRRP, addressing the crucial goals of dependability, cost-efficiency, and simplicity of maintenance in wireless sensor networks (Zang, et al., 2010). HBRRP is revealed as a viable approach for load balancing, network durability, reliability, and decreased routing overhead through a thorough study and simulation-based comparisons. This study makes strides in the design of WSN routing protocols by providing a solid substitute that has a number of advantages over the state-of-the-art methods (Zang, et al., 2010).
Title & published at	*"Dynamic Routing Scheme for Linking Wireless Sensor Network towards Internet of Things,"* 2021 5th International Conference on Electrical Information and Communication Technology (EICT)
Authors	S. J, S. Ajmani, U. P. K, D. Chouhon and D. K. S. M
Objective	The goal of this study is to enhance wireless sensor networks as an effective Internet of Things (IoT) solution. It focuses on the crucial requirement for dependable routing protocols in expansive IoT applications. While effective in addressing network requirements, current routing protocols frequently fall short in balancing energy efficiency and security limitations, which can negatively impact network performance as a whole (J, et al., 2021).
Contribution	This paper presents a novel routing method that prioritizes security and energy efficiency as a solution to these problems. The suggested approach makes use of a designated controller node to enable real-time monitoring of node cooperation, energy levels, and the identification of malicious behaviour inside the data transmission path (J, et al., 2021). The controller node actively monitors data flow throughout the path to spot problems caused by transmission errors, traffic jams, or malicious activity. Then, in cooperation with other nodes, it starts path recovery via alternative multipath routes, ultimately improving network resilience (J, et al., 2021).
Findings	The study's findings show that the safe and energy-efficient routing system that has been presented successfully overcomes the routing restrictions that are common in IoT applications. The plan successfully increases network dependability by proactively detecting and addressing problems, including transmission errors and hostile activities. Additionally, the inclusion of energy-saving measures guarantees a balanced strategy, easing possible worries about energy depletion (J, et al., 2021).
Conclusion	In conclusion, the development of wireless sensor networks for Internet of Things applications has advanced significantly. By proactively resolving routing restrictions, the suggested routing strategy, which was created to strike a compromise between security and energy efficiency, proves its power to improve network resilience (J, et al., 2021). This work can potentially enhance the general performance and viability of large-scale IoT networks, thereby promoting the development and success of the wider IoT ecosystem (J, et al., 2021).
Title & published at	*"Wireless Sensor Network Routing Algorithm Based on Traffic Prediction for Internet of Thing,"* 2021 International Conference on Electronic Communications, Internet of Things and Big Data (ICEIB)
Authors	B. Wang
Objective	This study attempts to use cutting-edge technologies like electronic communications, the Internet of Things (IoT), and Big Data to harness the tremendous economic growth occurring within wireless sensor networks (WSNs). The study responds by introducing a brand-new traffic prediction-based WSN routing algorithm (TP-WSNRA) designed to meet the requirements of the IoT environment (Wang, 2021).
Contribution	To improve WSN routing for IoT applications, the proposed TP-WSNRA algorithm makes use of clustering strategies and neural network (NN) prediction models (Wang, 2021). The algorithm improves path quality through cluster head selection, hence reducing possible congestion, by utilizing NN models to forecast future traffic loads. Additionally, nodes carefully select their next-hop connections, enhancing network efficiency and preventing congestion (Wang, 2021).
Findings	Results from simulations confirm the TP-WSNRA algorithm's effectiveness. The suggested method shows a significant 10% increase in packet delivery rate over algorithms without traffic prediction capabilities. Additionally, a 19% reduction in the typical end-to-end delay indicates improved network responsiveness and effectiveness (Wang, 2021).
Conclusion	In conclusion, this study discusses how the development of IoT and other cutting-edge technology developments are driving changes in the wireless sensor network environment. By utilizing traffic load estimates and neural network models, the TP-WSNRA algorithm presents a predictive strategy that improves routing within WSNs. The outcomes show that the algorithm has a great deal of potential to improve network performance, decrease delays, and increase the overall efficacy and efficiency of WSNs in IoT contexts.
Title & published at	*"IoT-RF: A routing framework for the Internet of Things,"* 2017 IEEE 28th Annual International Symposium on Personal, Indoor, and Mobile Radio Communications (PIMRC)
Authors	M. O. Farooq and T. Kunz
Objective	In the context of the Internet of Things (IoT), this paper discusses the performance restrictions of the Routing Protocol for Low-Power and Lossy Networks (RPL) standardized by the Internet Engineering Task Force (IETF) in many-to-many and one-to-many communications. In particular, it addresses concerns about control overhead and the non-storing mode of operation (MoP) inherent to RPL. It presents a novel routing framework for IoT (IoT-RF) designed to handle many-to-one, many-to-many, and one-to-many communications (Farooq & Kunz, 2017).
Contribution	The IoT-RF framework described in this paper promotes the usage of a storing MoP and advises the installation of several gateways inside a network to get around the storage-related memory constraints. Additionally, it presents two novel protocols for identifying one-to-many and many-to-many forwarding pathways. To find these paths, one protocol uses many-to-one communication, which reduces additional control overhead. For pathfinding, the other protocol uses empty control messages (Farooq & Kunz, 2017).

Findings	Through Cooja-based emulation studies, the performance of the IoT-RF implementation within the Contiki operating system is compared to that of a widely used IoT routing framework. Results show that IoT-RF significantly outperforms the de facto framework in terms of packet delivery ratio, end-to-end packet delivery delay, and control overhead across many-to-one, many-to-many, and one-to-many communication situations(Farooq & Kunz, 2017).
Conclusion	In conclusion, this study introduces IoT-RF as a cutting-edge routing framework for the Internet of Things (IoT) that successfully tackles the drawbacks of existing solutions, particularly RPL, in many-to-many and one-to-many communications. IoT-RF significantly reduces packet delivery, latency, and control overhead by promoting MoP storage and providing novel path discovery protocols. This study offers insightful analysis and useful improvements to the field of IoT routing protocols, with potential use in numerous IoT scenarios (Farooq & Kunz, 2017).
Title & published at	*'Routing Protocols for Wireless Sensor Networks (WSNs)'*, Wireless Sensor Networks - Insights and Innovations. InTech, Oct. 04, 2017
Authors	N. Shabbir and S. R. Hassan
Objective	The expanding significance of wireless sensor networks (WSNs) and the necessity of customizing routing protocols and network architectures to fit particular application requirements are both highlighted by this study. Despite the widespread use of WSNs, not all applications necessitate quick data transfer with little pause; some favor throughput over delay. This study attempts to shed light on how to choose the best routing protocols and network layouts depending on application preferences. To do this, the study uses WLAN- and ZigBee-based Wireless Sensor Networks to do a performance analysis of several routing protocols (Shabbir & Hassan, 2017).
Contribution	The study makes a contribution by emphasizing the variety of application needs found in wireless sensor networks and the significance of matching routing protocols and network designs appropriately. The study provides helpful advice for choosing the best strategy based on preferences for certain applications by conducting a comparative performance analysis across several routing protocols. Given the various capabilities and limitations of WLAN and ZigBee-based networks, the inclusion of both networks deepens the analysis (Shabbir & Hassan, 2017).
Findings	The results of the study shed light on the trade-offs that occur in wireless sensor networks between data transit speed and interruption tolerance (Shabbir & Hassan, 2017). The research clarifies whether routing methods are more appropriate for particular application scenarios by contrasting the performance of different routing protocols. The analysis's usage of WLAN and ZigBee-based networks enables a thorough assessment of routing protocol performance in various technical scenarios (Shabbir & Hassan, 2017).
Conclusion	This study's findings highlight the expanding significance of wireless sensor networks across a range of applications. It underlines the necessity of matching network topologies and routing protocols to particular application requirements, whether throughput is prioritized or delay is reduced. The analysis of WLAN and ZigBee-based WSN performance offers helpful recommendations for choosing the best technology stack for various use cases, eventually enhancing the performance and effectiveness of WSN deployments (Shabbir & Hassan, 2017).
Title & published at	*"Analysis of the routing methods and Internet models for IoT,"* 2019 International Conference on Information Science and Communications Technologies (ICISCT)
Authors	J. Khoshimov
Objective	In the context of the TCP/IP model, this study undertakes an extensive analysis of several routing protocols. The study conducts a comparative evaluation based on important performance indicators, such as packet delivery ratio and throughput, across various network conditions using the NS-3 simulation environment (Khoshimov, 2019). Variations in the quantity of static and mobile nodes as well as various power levels are included in these situations. The research finds a common problem by analysing the underlying problems with existing protocols: these protocols were not initially created for the present-day Internet and the Internet of Things (IoT). The report promotes the use of the Named Data Networking (NDN) model as a suggested remedy and outlines its benefits in the context of the IoT (Khoshimov, 2019).
Contribution	This study makes a contribution by offering a thorough evaluation of the TCP/IP model's routing protocols' functionality in various network scenarios. It clarifies the common issues caused by TCP/IP's antiquated design, which is insufficient for the needs of modern Internet and IoT applications. In addition, the paper promotes the use of the NDN model as a potential remedy, explaining its benefits in the IoT environment (Khoshimov, 2019).
Findings	This study identifies widespread issues with existing routing protocols that result from those systems' TCP/IP roots through meticulous simulation studies. These problems continue across a variety of network setups with variable node counts and power levels. The paper also highlights the merits of the NDN model as a superior substitute for IoT applications, addressing the drawbacks of the conventional TCP/IP architecture (Khoshimov, 2019).
Conclusion	It proves that the TCP/IP-based architecture of existing protocols is the cause of their shortcomings. The research recommends using the NDN paradigm as a futuristic fix, demonstrating its applicability and benefits inside the IoT area (Khoshimov, 2019).

3. SYSTEMATIC ANALYSIS OF PERFORMANCE IN WSN

The systematic analysis of the routing protocols is divided into 2 different sections. The first part of the performance analysis deals with the performance of purely

Figure 1. Comparison of routing protocols of WSN
Source: Shabbir and Hassan (2017)

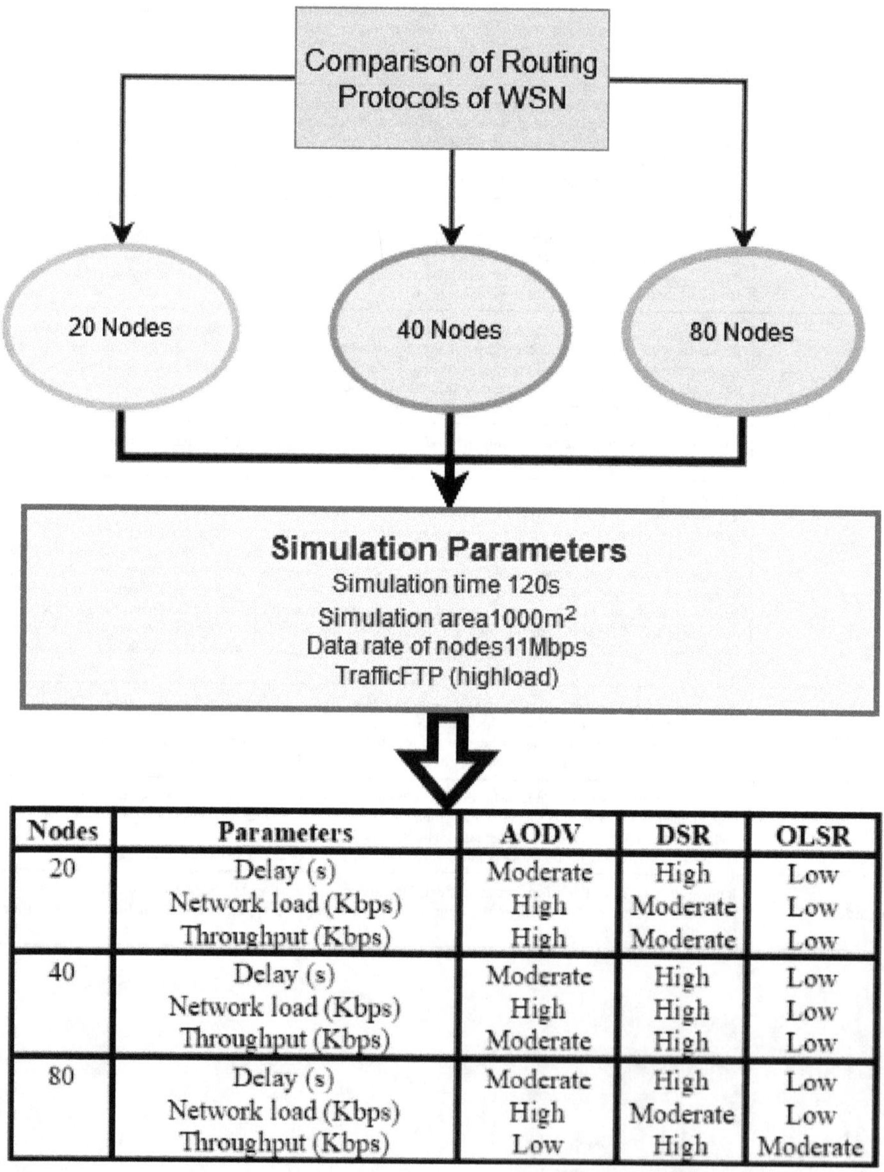

Nodes	Parameters	AODV	DSR	OLSR
20	Delay (s) Network load (Kbps) Throughput (Kbps)	Moderate High High	High Moderate Moderate	Low Low Low
40	Delay (s) Network load (Kbps) Throughput (Kbps)	Moderate High Moderate	High High High	Low Low Low
80	Delay (s) Network load (Kbps) Throughput (Kbps)	Moderate High Low	High Moderate High	Low Low Moderate

Figure 2. Simulation model

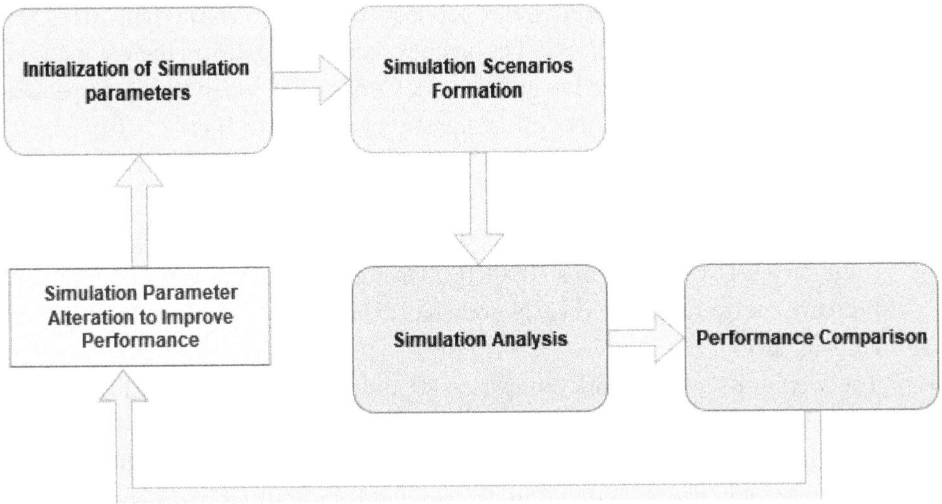

proactive or reactive routing protocols, and the second part deals with the performance of cluster-cum-routing methods.

3.1. Performance Analysis of Routing Protocol

The comparative analysis of the Ad Hoc on-demand distance vector (AODV) routing protocol, optimized link state routing (OLSR) protocol, and dynamic source routing (DSR) protocol is shown in Figure 1.

To examine the AODV, DSR, and OLSR routing protocols in WLAN-based WSNs, the authors analyzed the performance using the OPNET Modeler 14.5 network simulator. Prior studies have shown that these protocols work better and are compatible with WLAN-based WSNs. These protocols' performance in terms of delay, throughput, and network load will be assessed in small, medium, and large-scale networks. Twenty nodes make up a small-size network, forty for a medium-scale network, and eighty for a large-scale network.

- Three network metrics are defined, i.e., End-to-End delay, throughput, and network load (Shabbir & Hassan, 2017).
- End-to-end (ETE) delay measures the time it takes for data to travel from source to destination through a network, encompassing retransmission delays, packet transfer time, broadcast delay, and other route-related delays (Shabbir & Hassan, 2017).

- Throughput quantifies the data transmitted from source to destination within a specified time, measured in bytes per second (Shabbir & Hassan, 2017).
- Network load (NL) is expressed in bits per second. It signifies the net load on the network, often referred to as Network Congestion, when traffic surpasses link capacity, leading to network congestion (Shabbir & Hassan, 2017).
- In simulations, the authors have considered sensor networks on small, medium, and large scales.
- The small-scale network has 20 nodes connected in a star topology with a stationary WLAN server in a 1000x1000 m area, using IPv4 and File Transfer Protocol for traffic. Each WLAN node has a data rate of 11 Mbps (Shabbir & Hassan, 2017).
- The medium-scale network comprises 40 nodes, and the large-scale network consists of 80 nodes (Shabbir & Hassan, 2017).
- The simulation setup is shown in Figure 2. After running simulations, the results are analyzed for delay in seconds and throughput and network load in bits per second for AODV in small, medium, and large-scale networks, respectively (Shabbir & Hassan, 2017).

The Figure 1 lists the outcomes of small, medium, and large-size networks. It can be concluded that OLSR performs significantly (>50%) better in extensive networks than the other two protocols regarding delay. In contrast, AODV performs significantly (>100%) better in small and medium-sized networks. In all three circumstances, OLSR provides the lowest load possible for the network (Shabbir & Hassan, 2017).

The best throughput in small-scale networks is provided by AODV, which is 86% more than OLSR and 40% more than DSR. In a medium-scale network, DSR outperforms AODV and OLSR by a ratio of 13 and 40%, respectively. Like small networks, massive networks outperform them by a factor of 47 and 18% (Shabbir & Hassan, 2017).

The same comparison applies to an AODV-based ZigBee wireless sensor network. ZIGBEE nodes are used in applications requiring lower data rates and longer battery life. Higher data speeds are provided via wireless sensor nodes. However, this has the drawback that they use more power. Therefore, the author employed ZIGBEE in those applications where we don't need more excellent data rates because they prolong the network's life (Shabbir & Hassan, 2017).

The analysis shows that in a network with ZIGBEE nodes, the end-to-end delay is more significant. End-to-end delay begins at 0.060 s, steps up initially, and eventually reaches saturation at about 0.070 s. Compared to WSN nodes, a ZIGBEE network has lesser throughput and End-to-End delays that barely get over 0.010 seconds (Shabbir & Hassan, 2017). As per the results, throughput initially rises linearly before stabilizing at 6300 bits/s. Because ZIGBEE is a low-power, low-cost technology,

Figure 3. Comparison of cluster-cum-routing protocols of WSN

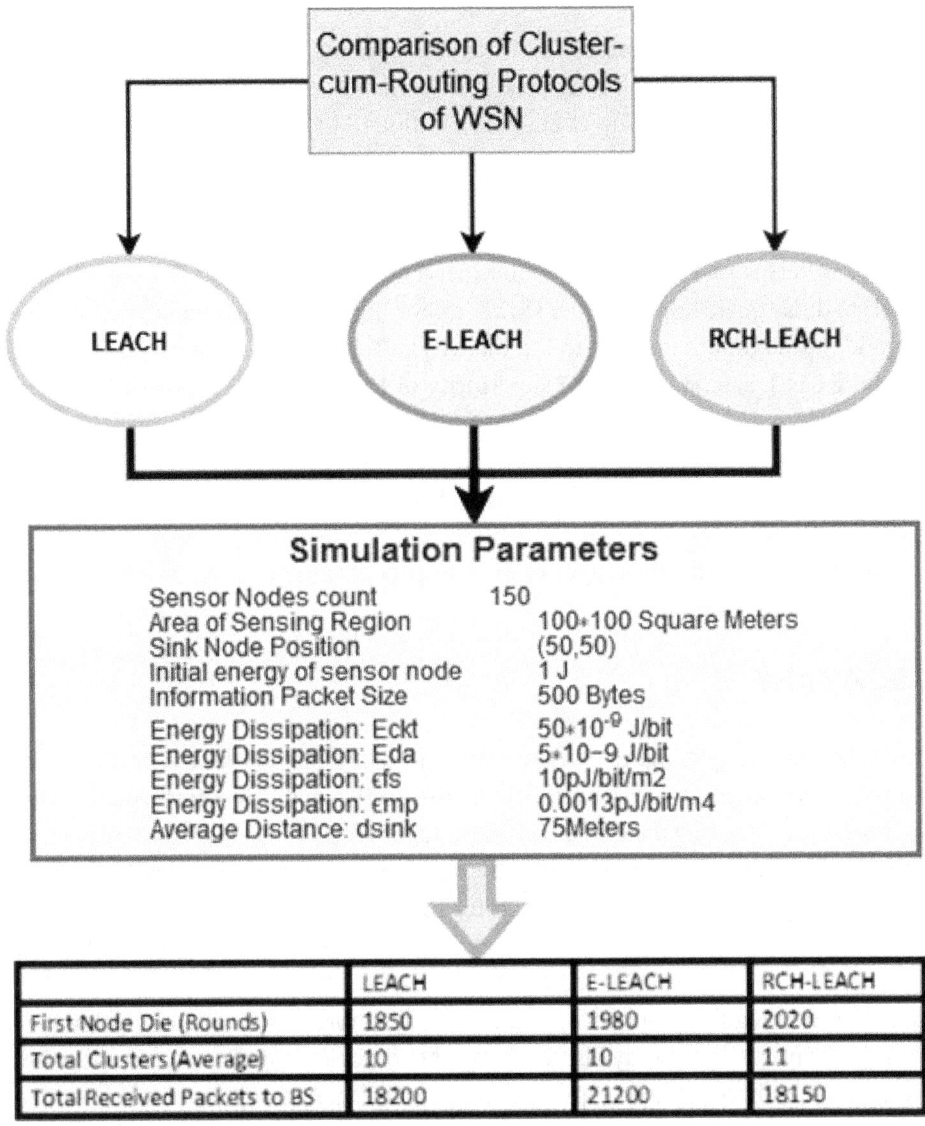

	LEACH	E-LEACH	RCH-LEACH
First Node Die (Rounds)	1850	1980	2020
Total Clusters (Average)	10	10	11
Total Received Packets to BS	18200	21200	18150

ZIGBEE nodes are employed when there are worries about the network's lifespan and financial issues (Shabbir & Hassan, 2017).

3.2 Performance Analysis of Cluster-cum-Routing Protocol

In this subsection, the performance of LEACH is compared and analyzed with the other existing cluster-cum-routing protocols (Panchal, et al., 2020; Vellaichamy, et al., 2023; Mirjalili, 2016; Guo et al., 2023; Patidar, et al., 2022a).

Figure 3 shows a comparative analysis of the LEACH, E-LEACH, and RCH-LEACH (Panchal, et al., 2020) protocols. The parameters are also shown in Figure 3. Panchal et al. (2020) reported the performance improvement of the RCH-LEACH protocol in terms of the network's lifetime. The network performs optimally for moderate data traffic applications. However, for high-data traffic applications, the network's throughput is not up to the mark. The first node dies, and the last node dies, for RCH-LEACH proves the superiority of RCH-LEACH protocol.

4. ADVANTAGE AND DISADVANTAGE OF VARIOUS ASPECTS

The advantages and disadvantages of various aspects are given below in Table 2.

5. SUMMARY

In this chapter, the systematic literature study is accomplished for the performance analysis of the various routing protocols. In this study, analysis is performed for the two categories of routing methods. The first category is proactive and reactive routing methods, generally used for ad hoc networks. The second category is clustering-centric routing methods, generally used for the wireless sensor network. Both the different categories of the protocol are helpful for the Internet of Things (IoT). The research also discusses the manuscripts related to several opportunistic routing algorithms that regularly outperform conventional shortest-path routing techniques in various contexts. The study also discusses how the development of IoT and other cutting-edge technology developments are driving changes in the wireless sensor network environment. In the end, the systematic analysis of the results shows that different protocols perform suitably for different applications. Therefore, we can conclude that selecting a suitable protocol depends on the application requirement.

6. FUTURE SCOPE

There are several exciting directions for further research and development in the rapidly developing fields of Wireless Sensor Networks (WSN) and the Internet of

Table 2. Disadvantages and advantages of various aspects

Aspect	Advantages	Disadvantages
Energy Efficiency	Advantages: Energy-efficient protocols can significantly prolong the battery life of sensor nodes, making them suitable for long-term deployments.	Disadvantages: Implementing energy-efficient protocols may require complex algorithms and can introduce overhead, impacting network performance.
Data Transmission Efficiency	Advantages: Efficient data transmission ensures timely and reliable delivery of data, critical for real-time applications.	Disadvantages: Overhead from routing and clustering protocols may lead to increased latency and reduced network throughput.
Application Adaptability	Advantages: Protocols can be tailored to suit specific application requirements, allowing customization for different use cases.	Disadvantages: The need for application-specific customization may complicate the deployment and management of WSNs and IoT networks.
Scalability and Network Size	Advantages: Protocols designed for scalability can accommodate network growth without sacrificing performance.	Disadvantages: Scalability may introduce complexity and challenges in maintaining network stability, especially in dynamic environments.
Security and Privacy	Advantages: Secure routing and clustering protocols protect data from unauthorized access and ensure privacy in sensitive applications.	Disadvantages: Implementing security measures can be resource-intensive and may add complexity to the network.
Cross-Layer Integration	Advantages: Cross-layer optimization can enhance network performance by coordinating different protocol layers.	Disadvantages: Complex interactions between layers may introduce challenges in protocol design and maintenance.
Machine Learning Integration	Advantages: AI and ML integration can enable intelligent decision-making, self-optimization, and anomaly detection.	Disadvantages: Implementing AI and ML may require significant computational resources and expertise.
Hybrid Networking (e.g., 5G Integration)	Advantages: Integrating with 5G networks can provide high data rates and low latency, enabling new IoT applications.	Disadvantages: Hybrid networking can be technically challenging and may require substantial infrastructure investments.
Blockchain Integration	Advantages: Blockchain enhances data integrity, traceability, and security, crucial in applications like supply chain and healthcare.	Disadvantages: Blockchain implementation can introduce complexity, higher processing overhead, and latency.

Things (IoT). One possible path is the integration of artificial intelligence (AI) and machine learning (ML) into routing and clustering protocols. AI and ML can be used by WSNs and IoT devices to enhance predictive analytics, make intelligent, and adjust to changing environmental conditions. One method to achieve this is to develop self-learning routing algorithms that can foresee mistakes, enhance network performance, and proactively handle issues. Moreover, machine learning can help with anomaly detection, which is crucial for identifying and managing network anomalies and security breaches. By using AI, WSNs and IoT can become more efficient, adaptive, and secure. The efficiency, adaptability, and security of WSNs and IoT can be increased by integrating AI and ML into these protocols.

Furthermore, as IoT applications continue to diversify, there is an increasing need for hybrid routing and clustering protocols that can operate smoothly with networks that are beyond 5G. The advent of 5G technology results in lower latency, quicker data speeds, and increased network density. Future research may focus on developing hybrid protocols that combine the benefits of WSNs and 5G networks to provide efficient and reliable data transport. This could lead to new applications where 5G connectivity is used to work with WSNs and IoT devices to meet the demanding requirements of industries like industrial automation, smart grids, and autonomous vehicles. Additionally, integrating blockchain technology into IoT networks to provide safe and transparent data transfers is an exciting new direction to pursue. Applications like supply chain management and healthcare can benefit greatly from blockchain's capacity to improve data integrity, traceability, and security. All things considered, there is a great deal of room for innovation in routing and clustering protocols in WSNs and IoT, particularly when combining blockchain, AI, ML, and hybrid networking.

REFERENCES

Abbasi, A. A., & Younis, M. (2007). A survey on clustering algorithms for wireless sensor networks. *Computer Communications*, *30*(14-15), 2826–2841. doi:10.1016/j.comcom.2007.05.024

Agbehadji, I. E., Millham, R. C., Fong, S. J., Jung, J. J., Bui, K.-H. N., Abayomi, A., & Frimpong, S. O. (2019). Bio-inspired energy efficient clustering approach for wireless sensor networks. In *2019 International Conference on Wireless Networks and Mobile Communications (WINCOM)* (pp. 1-8). Fez, Morocco: IEEE. 10.1109/WINCOM47513.2019.8942532

Behera, T. M., Mohapatra, S. K., Samal, U. C., Khan, M. S., Daneshmand, M., & Gandomi, A. H. (2019, June). Residual Energy-Based Cluster-Head Selection in WSNs for IoT Application. *IEEE Internet of Things Journal*, *6*(3), 5132–5139. doi:10.1109/JIOT.2019.2897119

Clare, L. P., Pottie, G. J., & Agre, J. R. (1999). *Self-organizing distributed sensor networks. Defense, Security, and Sensing Unattended Ground Sensor Technologies and Applications*. SPIE. doi:10.1117/12.357138

Ding, P., Holliday, J., & Celik, A. (2005). *Distributed Energy-Efficient Hierarchical Clustering for Wireless Sensor Networks. In Distributed Computing in Sensor Systems DCOSS 2005*. Springer., doi:10.1007/11502593_25

Farooq, M. O., & Kunz, T. (2017). IoT-RF: A routing framework for the Internet of Things. In *2017 IEEE 28th Annual International Symposium on Personal, Indoor, and Mobile Radio Communications (PIMRC)* (pp. 1-7). IEEE. 10.1109/PIMRC.2017.8292730

Fradj, H. B., Anane, R., & Bouallegue, R. (2018). Energy Consumption for Opportunistic Routing Algorithms in WSN. In *2018 IEEE 32nd International Conference on Advanced Information Networking and Applications (AINA)* (pp. 259-265). IEEE. 10.1109/AINA.2018.00048

Guo, X., Ye, Y., Li, L., Wu, R., & Sun, X. (2023, March 03). WSN Clustering Routing Algorithm Combining Sine Cosine Algorithm and Lévy Mutation. *IEEE Access : Practical Innovations, Open Solutions*, *11*, 22654–22663. doi:10.1109/ACCESS.2023.3252027

Handy, M., Haase, M., & Timmermann, D. (2002). Low energy adaptive clustering hierarchy with deterministic cluster-head selection. In *4th International Workshop on Mobile and Wireless Communications Network* (pp. 368-372). IEEE. 10.1109/MWCN.2002.1045790

Heinzelman, W., Chandrakasan, A., & Balakrishnan, H. (2000). Energy-efficient communication protocol for wireless microsensor networks. In *Proceedings of the 33rd Annual Hawaii International Conference on System Sciences* (pp. 1-10). IEEE. 10.1109/HICSS.2000.926982

Hoang, D., Kumar, R., & Panda, S. (2010). Fuzzy C-Means clustering protocol for Wireless Sensor Networks. In *2010 IEEE International Symposium on Industrial Electronics* (pp. 3477-3482). IEEE. 10.1109/ISIE.2010.5637779

J, S., Ajmani, S., K, U. P., Chouhon, D., & M, D. K. (2021). Dynamic Routing Scheme for Linking Wireless Sensor Network towards Internet of Things. In *2021 5th International Conference on Electrical Information and Communication Technology (EICT)* (pp. 1-4). IEEE. doi:10.1109/EICT54103.2021.9733501

Javidi, A., & Buhler, E. V. (2016). *Opportunistic Routing in Wireless Networks.* Now Publishers. doi:10.1561/1300000021

Khoshimov, J. (2019). Analysis of the routing methods and Internet models for IoT. In *2019 International Conference on Information Science and Communications Technologies (ICISCT)* (pp. 1-5). IEEE. 10.1109/ICISCT47635.2019.9011968

Kumar, P., Dwivedi, R., & Tyagi, V. (2019). Fuzzy Ant Colony Optimization Based Energy Efficient Routing For Mixed Wireless Sensor Network. In *2019 International Conference on Issues and Challenges in Intelligent Computing Techniques (ICICT)* (pp. 1-7). IEEE. 10.1109/ICICT46931.2019.8977699

Leu, J.-S., Chiang, T.-H., Yu, M.-C., & Su, K.-W. (2015). Feburary). Energy Efficient Clustering Scheme for Prolonging the Lifetime of Wireless Sensor Network With Isolated Nodes. *IEEE Communications Letters, 19*(2), 259–262. doi:10.1109/LCOMM.2014.2379715

Lotf, J. J., Bonab, M. N., & Khorsandi, S. (2008). A Novel cluster-based routing protocol with extending lifetime for wireless sensor networks. In *2008 5th IFIP International Conference on Wireless and Optical Communications Networks (WOCN '08)* (pp. 1-5). IEEE. 10.1109/WOCN.2008.4542499

M, E., RS, R., M, H., K, S., & O, A. (2020). Swarm intelligence–based energy efficient clustering with multihop routing protocol for sustainable wireless sensor networks. *International Journal of Distributed Sensor Networks, 16*(9), 1-12. doi:10.1177/1550147720949133

Mabrouki, J., Azrour, M., Dhiba, D., Farhaoui, Y., & Hajjaji, S. E. (2021, March). IoT-based data logger for weather monitoring using arduino-based wireless sensor networks with remote graphical application and alerts. *Big Data Mining and Analytics, 4*(1), 25–32. doi:10.26599/BDMA.2020.9020018

Mirjalili, S. (2016). SCA: A Sine Cosine Algorithm for solving optimization problems. *Knowledge-Based Systems, 96*, 120–133. doi:10.1016/j.knosys.2015.12.022

Mody, S., Mirkar, S., Ghag, R., & Kotecha, P. (2021). Cluster Head Selection Algorithm For Wireless Sensor Networks Using Machine Learning. In *2021 International Conference on Computational Performance Evaluation (ComPE)* (pp. 445-450). IEEE. 10.1109/ComPE53109.2021.9752264

Panchal, A., Singh, L., & Singh, R. K. (2020). RCH-LEACH: Residual Energy based Cluster Head Selection in LEACH for Wireless Sensor Networks. In *2020 International Conference on Electrical and Electronics Engineering (ICE3)* (pp. 322-325). IEEE. 10.1109/ICE348803.2020.9122962

Patidar, Y., Jain, M., & Vyas, A. K. (2022a). Comparative Analysis of Clustering and Routing Protocols for Energy-Efficient WSN. In *2022 4th International Conference on Advances in Computing, Communication Control and Networking (ICAC3N)* (pp. 1768-1773). IEEE. 10.1109/ICAC3N56670.2022.10074074

Patidar, Y., Jain, M., & Vyas, A. K. (2022b). Modified Unexpended Energy Based Stable Cluster Head Selection for Homogeneous Wireless Sensor Networks (WSNs). In *2022 3rd International Conference on Computing, Analytics and Networks (ICAN)* (pp. 1-5). IEEE. 10.1109/ICAN56228.2022.10007098

Saranya, V., Shankar, S., & Kanagachidambaresan, G. R. (2018). Energy Efficient Clustering Scheme (EECS) for Wireless Sensor Network with Mobile Sink. *Wireless Personal Communications*, *100*(4), 1553–1567. doi:10.1007/s11277-018-5653-1

Shabbir, N., & Hassan, S. R. (2017). Routing Protocols for Wireless Sensor Networks (WSNs). In Wireless Sensor Networks - Insights and Innovations. IntechOpen. doi:10.5772/intechopen.70208

Sharawi, M., & Emary, E. (2017). Impact of grey wolf optimization on WSN cluster formation and lifetime expansion. In *2017 Ninth International Conference on Advanced Computational Intelligence (ICACI)* (pp. 157-162). IEEE. 10.1109/ICACI.2017.7974501

Sun, W., Song, Y., & Chen, M. (2010). A load-balanced and energy-aware routing metric for wireless multimedia sensor networks. In *IET 3rd International Conference on Wireless, Mobile and Multimedia Networks (ICWMNN 2010)* (pp. 21-24). IET. 10.1049/cp.2010.0608

Tsai, T.-H., Huang, C.-C., Chang, C.-H., & Hussain, M. A. (2020). Design of Wireless Vision Sensor Network for Smart Home. *IEEE Access : Practical Innovations, Open Solutions*, *8*, 60455–60467. doi:10.1109/ACCESS.2020.2982438

Vellaichamy, J., Basheer, S., Bai, P., Khan, M., Kumar Mathivanan, S., Jayagopal, P., & Babu, J. (2023). Wireless Sensor Networks Based on Multi-Criteria Clustering and Optimal Bio-Inspired Algorithm for Energy-Efficient Routing. *Applied Sciences (Basel, Switzerland)*, *13*(5), 1–14. doi:10.3390/app13052801

Wang, B. (2021). Wireless Sensor Network Routing Algorithm Based on Traffic Prediction for Internet of Thing. In *2021 International Conference on Electronic Communications, Internet of Things and Big Data (ICEIB)* (pp. 82-86). IEEE. 10.1109/ICEIB53692.2021.9686397

Xu, F., Ye, H., Yang, F., & Zhao, C. (2019, January 04). Software Defined Mission-Critical Wireless Sensor Network: Architecture and Edge Offloading Strategy. *IEEE Access : Practical Innovations, Open Solutions*, *7*, 10383–10391. doi:10.1109/ACCESS.2019.2890854

Younis, O., & Fahmy, S. (2004, October-December). HEED: A hybrid, energy-efficient, distributed clustering approach for ad hoc sensor networks. *IEEE Transactions on Mobile Computing*, *3*(5), 366–379. doi:10.1109/TMC.2004.41

Yousif, Y. K., Badlishah, R., Yaakob, N., & Amir, A. (2017). An Energy Efficient and Load Balancing Clustering Scheme for Wireless Sensor Network (WSN) Based on Distributed Approach. In *J. Phys.: Conf. Ser. 1st International Conference on Green and Sustainable Computing (ICoGeS)* (pp. 1-6). IOP Publishing Ltd. 10.1088/1742-6596/1019/1/012007

Zagrouba, R., & Kardi, A. (2021). Comparative Study of Energy Efficient Routing Techniques in Wireless Sensor Networks. *Information (Basel)*, *12*(1), 1–28. doi:10.3390/info12010042

Zang, Z., Qi, J.-D., & Cao, Y.-J. (2010). A Robust Routing Protocol in wireless sensor network. In *IET International Conference on Wireless Sensor Network 2010 (IET-WSN 2010)* (pp. 276-279). IET. 10.1049/cp.2010.1066

Zhixiang, D., & Bensheng, Q. (2007). Three-layered routing protocol for WSN based on LEACH algorithm. In *IET Conference on Wireless, Mobile and Sensor Networks 2007 (CCWMSN07)* (pp. 72-75). IET. 10.1049/cp:20070086

Compilation of References

Abbasi, A. A., & Younis, M. (2007). A survey on clustering algorithms for wireless sensor networks. *Computer Communications*, *30*(14-15), 2826–2841. doi:10.1016/j.comcom.2007.05.024

Abdelmoamen Ahmed, A., Al Omari, S., Awal, R., Fares, A., & Chouikha, M. (2021). A distributed system for supporting smart irrigation using internet of things technology. *Engineering Reports*, *3*(7), e12352. doi:10.1002/eng2.12352

AbdKadir. (2020). IoT based Car Parking Management System using IR Sensor. *Journal of Computing Research and Innovation*, *5*(2).

Adeniyi, E. A., Ogundokun, R. O., Misra, S., Awotunde, J. B., & Abiodun, K. M. (2022). Enhanced security and privacy issue in multi-tenant environment of green computing using blockchain technology. In *Blockchain Applications in the Smart Era* (pp. 65–83). Springer International Publishing. doi:10.1007/978-3-030-89546-4_4

Adi, E., Anwar, A., Baig, Z., & Zeadally, S. (2020). Machine learning and data analytics for the IoT. *Neural Computing & Applications*, *32*(20), 16205–16233. doi:10.1007/s00521-020-04874-y

Agbehadji, I. E., Millham, R. C., Fong, S. J., Jung, J. J., Bui, K.-H. N., Abayomi, A., & Frimpong, S. O. (2019). Bio-inspired energy efficient clustering approach for wireless sensor networks. In *2019 International Conference on Wireless Networks and Mobile Communications (WINCOM)* (pp. 1-8). Fez, Morocco: IEEE. 10.1109/WINCOM47513.2019.8942532

Agi, M. A., & Jha, A. K. (2022). Blockchain technology in the supply chain: An integrated theoretical perspective of organizational adoption. *International Journal of Production Economics*, *247*, 108458. doi:10.1016/j.ijpe.2022.108458

Ahmad. (n.d.). *Client server smart irrigation system*. Academic Press.

Ahmed, I., & Sikder, M. (2020). IoT based smart irrigation system at university of Chittagong, Bangladesh. *International Journal of Computer Applications*, *176*(15), 39–45. doi:10.5120/ijca2020920097

Ahmed, I., Zhang, Y., Jeon, G., Lin, W., Khosravi, M. R., & Qi, L. (2022). A blockchain-and artificial intelligence-enabled smart IoT framework for sustainable city. *International Journal of Intelligent Systems*, *37*(9), 6493–6507. doi:10.1002/int.22852

Ahmed, N., Michelin, R. A., Xue, W., Ruj, S., Malaney, R., Kanhere, S. S., Seneviratne, A., Hu, W., Janicke, H., & Jha, S. K. (2020). A survey of COVID-19 contact tracing apps. *IEEE Access : Practical Innovations, Open Solutions*, *8*, 134577–134601. doi:10.1109/ACCESS.2020.3010226

Aker, J. C., Ghosh, I., & Burrell, J. (2016, November). The promise (and pitfalls) of ICT for agriculture initiatives. *Agricultural Economics*, *47*(S1, no. S1), 35–48. doi:10.1111/agec.12301

Akter, S. (2018). Developing a smart irrigation system using arduino. *International Journal of Research Studies in Science, Engineering and Technology*, *6*(1), 31–39.

Al Nahian. (2021). *IoT based soil monitoring and automatic irrigation system.* Academic Press.

Albawi, S., Mohammed, T. A., & Al-Zawi, S. (2017). Understanding of a convolutional neural network. 2017 international conference on engineering and technology (ICET), 1–6. doi:10.1109/ICEngTechnol.2017.8308186

Al-Khayari, A., Al-Khayari, H., Al-Nabhani, S., Bait-Suwailam, M., & Nadir, Z. (2013). *Design of an Enhanced RF Energy Harvesting System for Wireless Sensors.* IEEE GCC Conference And Exhibition. doi:10.1109/IEEEGCC.2013.6705826

Al-Lawati, M., Al-Busaidi, M., & Nadir, Z. (2012). *RF Energy Harvesting System Design for Wireless Sensors.* International Multi-Conference on Systems, Signals, and Devices. doi:10.1109/SSD.2012.6197969

Allee, G. (n.d.). *Smart CITIES USA.* Smart America. Retrieved from https://smartamerica.org/teams/smart-cities-usa/

Alromaihi, N., Ismail, Y., & Elmedany, W. (2022, October). Literature Review of Blockchain-based Cloud Computing: Data Security Issues and Challenges. In *2022 International Conference on Data Analytics for Business and Industry (ICDABI)* (pp. 484-492). IEEE. 10.1109/ICDABI56818.2022.10041637

Al-Turjman, F., & Salama, R. (2020). An Overview about the Cyberattacks in Grid and Like Systems. *Smart Grid in IoT-Enabled Spaces*, 233-247.

Al-Turjman, F., & Salama, R. (2021a). Cyber security in mobile social networks. In Security in IoT social networks (pp. 55-81). Academic Press.

Al-Turjman, F., & Salama, R. (2021b). Security in social networks. In *Security in IoT Social Networks* (pp. 1–27). Academic Press.

Anwita, Shrey, Yelithoti, Tapaswini, & Swati. (2021). Smart Parking System Based on the IR Sensor And Node MCU with the Blynk Application. *International Journal of Sensors, Wireless Communications and Control, 11*(4).

Arjunan, S., & Sujatha, P. (2018). Lifetime maximization of wireless sensor network using fuzzy based unequal clustering and ACO based routing hybrid protocol. *Applied Intelligence*, *48*(8), 2229–2246. doi:10.1007/s10489-017-1077-y

Arudra Annepu, M. (2021). IoT Enabled Smart Lighting System for Smart Cities. *Journal of Physics: Conference Series*, *2089*(1), 012037. doi:10.1088/1742-6596/2089/1/012037

Ashour, A. S., El-Attar, A., Dey, N., Abd El-Kader, H., & Abd El-Naby, M. M. (2020). Long short term memory based patient-dependent model for fog detection in parkinson's disease. *Pattern Recognition Letters*, *131*, 23–29. doi:10.1016/j.patrec.2019.11.036

Ashreetha, Anandaram, Gupta, & Verma. (2023). IoT Wearable Breast Temperature Assessment System. *2023 7th International Conference on Computing Methodologies and Communication (ICCMC)*, 1236-1241. 10.1109/ICCMC56507.2023.10083511

Atzori, L., Iera, A., & Morabito, G. (2010). The internet of things: A survey. *Computer Networks*, *54*(15), 2787–2805. doi:10.1016/j.comnet.2010.05.010

Avinash Sharma, Rao, Shankar, Sarma, Chaturvedi, & Hussain. (2022). Industrial quality healthcare services using Internet of Things and fog computing approach. *Measurement: Sensors*, *24*. doi:10.1016/j.measen.2022.100517

Awadallah, R., Samsudin, A., Teh, J. S., & Almazrooie, M. (2021). An integrated architecture for maintaining security in cloud computing based on blockchain. *IEEE Access : Practical Innovations, Open Solutions*, *9*, 69513–69526. doi:10.1109/ACCESS.2021.3077123

Ayaz, M., Ammad-Uddin, M., Sharif, Z., Mansour, A., & Aggoune, E.-H. M. (2019). Internet-of-things (iot)-based smart agriculture: Toward making the fields talk. *IEEE Access : Practical Innovations, Open Solutions*, *7*, 129551–129583. doi:10.1109/ACCESS.2019.2932609

Azshwanth, D. (2019). Automated car parking system. *International Conference on Physics and Photonics Processes in Nano Sciences*, 1362.

Babaa, S. E. (2020). Smart irrigation system using arduino with solar power. *International Journal of Engineering Research & Technology (Ahmedabad)*, *9*(05), 7.

Balaji, V. (2019). Smart irrigation system using iot and image processing. *International Journal of Engineering and Advanced Technology*, *8*(6s, no. 6S), 115–120. doi:10.35940/ijeat.F1024.0886S19

Banerjee, B., Joshi, S., Thoday-Kennedy, E., Pasam, R. K., Tibbits, J., Hayden, M., Spangenberg, G., & Kant, S. (2020). High-throughput Phenotyping Using Digital and Hyperspectral Imaging-derived Biomarkers for Genotypic Nitrogen Response. *Journal of Experimental Botany*, *71*(15), 4604–4615. doi:10.1093/jxb/eraa143 PMID:32185382

Baravade, D. (2019). Study paper on smart irrigation system. *International Journal of Engineering Sciences & Research Technology*, *8*(4), 99–102.

Behera, T. M., Mohapatra, S. K., Samal, U. C., Khan, M. S., Daneshmand, M., & Gandomi, A. H. (2019, June). Residual Energy-Based Cluster-Head Selection in WSNs for IoT Application. *IEEE Internet of Things Journal*, *6*(3), 5132–5139. doi:10.1109/JIOT.2019.2897119

Bhatti, G. (2018). Machine learning based localization in large-scale wireless sensor networks. *Sensors (Basel)*, *18*(12), E4179. doi:10.3390/s18124179 PMID:30487457

Bonde, D. J., Shende, R. S., Gaikwad, K. S., Kedari, A. S., & Bhokre, A. U. (2014). Automated car parking system commanded by Android application. *Proc. Int. Conf. Comput. Commun. Inform. (ICCCI),* 5, 1-4.

Bruning, B., Liu, H., Brien, C., Berger, B., Lewis, M., & Garnett, T. (2019). The Development of Hyperspectral Distribution Maps to Predict the Content and Distribution of Nitrogen and Water in Wheat (Triticum Aestivum). *Frontiers in Plant Science*, *10*, 1380. Advance online publication. doi:10.3389/fpls.2019.01380 PMID:31737009

Chaudhary, H., Bansal, P., & Valarmathi, B. (2017). Advanced Car parking system using Arduino. *4th International Conference on Advanced Computing and Communication System*, 1-5.

Chen, T., & Guestrin, C. (2016). Xgboost: A scalable tree boosting system. *Proceedings of the 22nd acm sigkdd international conference on knowledge discovery and data mining*, 785–794. 10.1145/2939672.2939785

Chinamuttevi, V. S., Prasad, K., Begum, A. Y., Tanguturi, R. C., & Ahmed, B. (2022). Implementation of an Early Warning System using Internet of Things and Rainfall Threshold. *2022 6th International Conference on Electronics, Communication and Aerospace Technology*, 491-496. 10.1109/ICECA55336.2022.10009211

Choeychuen, K. (2013). Automatic parking lot mapping for available parking space detection. *Proceedings of the 5th International Conference on Knowledge and Smart Technology (KST),* 117–121.

Cisco. (2016). *Cisco Visual Networking Index Predicts Near-Tripling of IP Traffic by 2020*. Retrieved from https://newsroom.cisco.com/press-release-content?type=press-release&articleId=1771211

Clare, L. P., Pottie, G. J., & Agre, J. R. (1999). *Self-organizing distributed sensor networks. Defense, Security, and Sensing Unattended Ground Sensor Technologies and Applications*. SPIE. doi:10.1117/12.357138

Corti, M., Gallina, P. M., Cavalli, D., & Cabassi, G. (2017). Hyperspectral Imaging of Spinach Canopy under Combined Water and Nitrogen Stress to Estimate Biomass, Water, and Nitrogen Content. *Biosystems Engineering*, *158*, 38–50. doi:10.1016/j.biosystemseng.2017.03.006

Cynthia, Priya, & Gopinath. (2018). IOT based Smart Parking Management System. *International Journal of Recent Technology and Engineering,* 7(4S), 374-379.

D'Amato, G., Avitabile, G., Coviello, G., & Talarico, C. (2016). Towards a Novel Architecture for Beam String of Active Phased-Array Antennas. *International Midwest Symposium on Circuits and Systems (MWSCAS).*

Dasgupta, A. (2019). Smart irrigation: Iot-based irrigation monitoring system. *Proceedings of international ethical hacking conference 2018,* 395–403. 10.1007/978-981-13-1544-2_32

Difallah, W. (2018b). Intelligent irrigation management system. Energy, 9(9). doi:10.14569/IJACSA.2018.090954

Difallah, W. (2018a). Design of a solar powered smart irrigation system (spsis) using wsn as an iot device. *Proceedings of the 2018 International Conference on Software Engineering and Information Management*, 124–128. 10.1145/3178461.3178482

Ding, P., Holliday, J., & Celik, A. (2005). *Distributed Energy-Efficient Hierarchical Clustering for Wireless Sensor Networks. In Distributed Computing in Sensor Systems DCOSS 2005.* Springer., doi:10.1007/11502593_25

Dreyfus, J., Audureau, E., Bohbot, Y., Coisne, A., Lavie-Badie, Y., Bouchery, M., Flagiello, M., Bazire, B., Eggenspieler, F., Viau, F., Riant, E., Mbaki, Y., Eyharts, D., Senage, T., Modine, T., Nicol, M., Doguet, F., Nguyen, V., Le Tourneau, T., ... Messika-Zeitoun, D. (2022). TRI-SCORE: A new risk score for in-hospital mortality prediction after isolated tricuspid valve surgery. *European Heart Journal*, *43*(7), 654–662. doi:10.1093/eurheartj/ehab679 PMID:34586392

Farooq, M. O., & Kunz, T. (2017). IoT-RF: A routing framework for the Internet of Things. In *2017 IEEE 28th Annual International Symposium on Personal, Indoor, and Mobile Radio Communications (PIMRC)* (pp. 1-7). IEEE. 10.1109/PIMRC.2017.8292730

Fateh, A. M., & Salarnejad, A. A. (2022). The Scope of Blockchain Technology: A Meta-Synthesis Study of Applications, Benefits, Challenges, and Related Technologies. *Information & Management*, *8*(1), 245–300.

Felini, C., Merenda, M., & Corte, F. G. D. (2014). Dynamic impedance matching network for RF energy harvesting systems. *IEEE RFID Technology and Applications Conference (RFID-TA)*. 10.1109/RFID-TA.2014.6934206

Ferreira, M. I. A., Barbieri, F. A., Moreno, V. C., Penedo, T., & Tavares, J. M. R. (2022). Machine learning models for parkinson's disease detection and stage classification based on spatial-temporal gait parameters. *Gait & Posture*, *98*, 49–55. doi:10.1016/j.gaitpost.2022.08.014 PMID:36049418

Fradj, H. B., Anane, R., & Bouallegue, R. (2018). Energy Consumption for Opportunistic Routing Algorithms in WSN. In *2018 IEEE 32nd International Conference on Advanced Information Networking and Applications (AINA)* (pp. 259-265). IEEE. 10.1109/AINA.2018.00048

Fu, L., Wang, B., Yuan, T., Chen, X., Ao, Y., Fitzpatrick, T., Li, P., Zhou, Y., Lin, Y., Duan, Q., Luo, G., Fan, S., Lu, Y., Feng, A., Zhan, Y., Liang, B., Cai, W., Zhang, L., Du, X., ... Zou, H. (2020). Clinical characteristics of coronavirus disease 2019 (COVID-19) in China: A systematic review and meta-analysis. *The Journal of Infection*, *80*(6), 656–665. doi:10.1016/j.jinf.2020.03.041 PMID:32283155

Gartner. (2015). *Gartner says 6.4 billion Connected "Things" Will Be in Use in 2016, Up 30 Percent From 2015*. Retrieved from https://www.gartner.com/newsroom/id/3165317

Geipel, J., Link, J., Wirwahn, J. A., & Claupein, W. (2016). A Programmable Aerial Multispectral Camera System for In-season Crop Biomass and Nitrogen Content Estimation. *Agriculture*, *6*(1), 4. doi:10.3390/agriculture6010004

Ghate, V., & Vijayakumar, V. (2018). Machine learning for data aggregation in WSN: A survey. *International Journal of Pure and Applied Mathematics*, *118*(24), 1–12.

Gladence, L. M., Anu, V. M., Rathna, R., & Brumancia, E. (2020). Recommender system for home automation using IoT and artificial intelligence. *Journal of Ambient Intelligence and Humanized Computing*, (0123456789). Advance online publication. doi:10.1007/s12652-020-01968-2

Goldstein, A., Fink, L., Meitin, A., Bohadana, S., Lutenberg, O., & Ravid, G. (2018). Applying machine learning on sensor data for irrigation recommendations: Revealing the agronomist's tacit knowledge. *Precision Agriculture*, *19*(3), 421–444. doi:10.1007/s11119-017-9527-4

Goumopoulos, C., O'Flynn, B., & Kameas, A. (2014). Automated zone-specific irrigation with wireless sensor/actuator network and adaptable decision support. *Computers and Electronics in Agriculture*, *105*, 20–33. doi:10.1016/j.compag.2014.03.012

Guo, X., Ye, Y., Li, L., Wu, R., & Sun, X. (2023, March 03). WSN Clustering Routing Algorithm Combining Sine Cosine Algorithm and Lévy Mutation. *IEEE Access : Practical Innovations, Open Solutions*, *11*, 22654–22663. doi:10.1109/ACCESS.2023.3252027

Habib, G., Sharma, S., Ibrahim, S., Ahmad, I., Qureshi, S., & Ishfaq, M. (2022). Blockchain Technology: Benefits, Challenges, Applications, and Integration of Blockchain Technology with Cloud Computing. *Future Internet*, *14*(11), 341. doi:10.3390/fi14110341

Ham, H.-S., Kim, H.-H., Kim, M.-S., & Choi, M.-J. (2014). Linear SVM-Based Android Malware Detection for Reliable IoT Services. *Journal of Applied Mathematics*, *2014*, 594501. doi:10.1155/2014/594501

Handy, M., Haase, M., & Timmermann, D. (2002). Low energy adaptive clustering hierarchy with deterministic cluster-head selection. In *4th International Workshop on Mobile and Wireless Communications Network* (pp. 368-372). IEEE. 10.1109/MWCN.2002.1045790

Han, R., Yan, Z., Liang, X., & Yang, L. T. (2022). How can incentive mechanisms and blockchain benefit with each other? a survey. *ACM Computing Surveys*, *55*(7), 1–38. doi:10.1145/3539604

Haque, M. A., Ahmad, S., Eljialy, A. E. M., Uddin, M. Y., & Sonal, D. (2023). Internet of Things (IoT) based Model for Water Management System. *2023 International Conference on Smart Computing and Application (ICSCA)*, 1-5. 10.1109/ICSCA57840.2023.10087587

Hassan, A. (2021). A wirelessly controlled robot-based smart irrigation system by exploiting arduino. *Journal of Robotics and Control*, *2*(1), 29–34.

Heinzelman, W., Chandrakasan, A., & Balakrishnan, H. (2000). Energy-efficient communication protocol for wireless microsensor networks. In *Proceedings of the 33rd Annual Hawaii International Conference on System Sciences* (pp. 1-10). IEEE. 10.1109/HICSS.2000.926982

Hendrawan, I. N. R. (2019). Fuzzy based internet of things irrigation system. *1st International Conference on Cybernetics and Intelligent System (ICORIS)*, 1, 146–150. 10.1109/ICORIS.2019.8874900

Herrero, H., Waylen, P., Southworth, J., Khatami, R., Yang, D., & Child, B. (2020). A Healthy Park Needs Healthy Vegetation: The Story of Gorongosa National Park in the 21st Century. *Remote Sensing (Basel)*, *12*(3), 476. doi:10.3390/rs12030476

Hoang, D., Kumar, R., & Panda, S. (2010). Fuzzy C-Means clustering protocol for Wireless Sensor Networks. In *2010 IEEE International Symposium on Industrial Electronics* (pp. 3477-3482). IEEE. 10.1109/ISIE.2010.5637779

Hoomod, H., & Jebur, T. (2018). Applying self-organizing map and modified radial based neural network for clustering and routing optimal path in wireless network. *Journal of Physics: Conference Series*, *1003*, 012040. doi:10.1088/1742-6596/1003/1/012040

Hou, H., Peng, S., Xu, J., Yang, S., & Mao, Z. (2012). *Seasonal variations of CH4 and N2O emissions in response to water management of paddy fields located in Southeast China*. Academic Press.

Hsu, C. Y., Yang, W., Parikh, R. V., Anderson, A. H., Chen, T. K., Cohen, D. L., He, J., Mohanty, M. J., Lash, J. P., Mills, K. T., Muiru, A. N., Parsa, A., Saunders, M. R., Shafi, T., Townsend, R. R., Waikar, S. S., Wang, J., Wolf, M., Tan, T. C., ... Go, A. S. (2021). Race, genetic ancestry, and estimating kidney function in CKD. *The New England Journal of Medicine*, *385*(19), 1750–1760. doi:10.1056/NEJMoa2103753 PMID:34554660

Hu, J., Hao, Z. C., & Hong, W. (2017). Design of a Wideband Quad-Polarization Reconfigurable Patch Antenna Array Using a Stacked Structure. *IEEE Transactions on Antennas and Propagation*, *65*(6), 3014–3023. doi:10.1109/TAP.2017.2695529

Hwang, S. (2017). Monitoring and Controlling System for an IoT Based Smart Home. *Int. J. Control Autom.*, *10*(2), 339–348. doi:10.14257/ijca.2017.10.2.28

Imteaj, A. (2016). Iot based autonomous percipient irrigation system using raspberry pi. *19th International Conference on Computer and Information Technology (ICCIT)*, 563–568. 10.1109/ICCITECHN.2016.7860260

Iqbal, T., & Qureshi, S. (2022). The survey: Text generation models in deep learning. *Journal of King Saud University. Computer and Information Sciences*, *34*(6), 2515–2528. doi:10.1016/j.jksuci.2020.04.001

Islam. (n.d.). *Iot based smart irrigation monitoring & controlling system in agriculture*. Academic Press.

Ismail, N. (2019). *Smart irrigation system based on internet of things (iot). Journal of Physics: Conference Series*, 1339.

J, S., Ajmani, S., K, U. P., Chouhon, D., & M, D. K. (2021). Dynamic Routing Scheme for Linking Wireless Sensor Network towards Internet of Things. In *2021 5th International Conference on Electrical Information and Communication Technology (EICT)* (pp. 1-4). IEEE. doi:10.1109/EICT54103.2021.9733501

Jain, S., & Ramesh, D. (2020). Machine Learning convergence for weather based crop selection. *2020 IEEE Int. Students' Conf. Electr. Electron. Comput. Sci. SCEECS.*

Javidi, A., & Buhler, E. V. (2016). *Opportunistic Routing in Wireless Networks*. Now Publishers. doi:10.1561/1300000021

Jayadevan, J. (2020). A novel architecture for internet of things in precision agriculture. *International Journal of Applied Engineering Research*, *15*(3), 204–211.

Jimenez-Buendıa, M., Soto-Valles, F., Blaya-Ros, P. J., Toledo-Moreo, A., Domingo-Miguel, R., & Torres-Sánchez, R. (2021). High-density wi-fi based sensor network for efficient irrigation management in precision agriculture. *Applied Sciences (Basel, Switzerland)*, *11*(4), 1628. doi:10.3390/app11041628

Ji-Yong, S., Xiao-Bo, Z., Jie-Wen, Z., Kai-Liang, W., Zheng-Wei, C., Xiao-Wei, H., De-Tao, Z., & Holmes, M. (2012). Nondestructive Diagnostics of Nitrogen Deficiency by Cucumber Leaf Chlorophyll Distribution Map Based on near Infrared Hyperspectral Imaging. *Scientia Horticulturae*, *138*, 190–197. doi:10.1016/j.scienta.2012.02.024

Johri, A., & Tripathi, A. (2019). Parkinson disease detection using deep neural networks. *2019 twelfth international conference on contemporary computing (IC3)*, 1–4.

Joshi, D. (2019). Smart irrigation system using iot for farmers. *International Journal of Emerging Technologies and Innovative Research*, *6*(1), 207–209.

Kamble & Dehankar. (2015). Review on Automatic Car Parking Indicator System. *Journal on Recent and Innovation Trends*.

Kang, S., Liang, Z., Pan, Y., Shi, P., & Zhang, J. (2000). Alternate furrow irrigation for maize production in an arid area. *Agricultural Water Management*, *45*(3), 267–274. doi:10.1016/S0378-3774(00)00072-X

Kannan, K., Kumar, N. S., Logith, E., Kumar, R. M., & Prakash, O. S. (2020). An internet of things based smart irrigation using solenoid valve. *International Journal of Recent Technology and Engineering*, *9*(1), 2018–2023. doi:10.35940/ijrte.A1979.059120

Kapitanova, K., & Son, S. H. (2012). Machine learning basics. Intell. Sens. Networks Integr. Sens. Networks, Signal Process. Mach. Learn., Ml, 3–29. doi:10.1201/b14300-2

Karar, M. E. (2021). *A pilot study of smart agricultural irrigation using unmanned aerial vehicles and iot-based cloud system*. arXiv preprint arXiv:2101.01851.

Karpagam, J. (2021). *7th International Conference on Advanced Computing and Communication Systems, ICACCS 2021*. Academic Press.

Karpagam, J. (2020). Smart irrigation system using iot. *6th International Conference on Advanced Computing and Communication Systems (ICACCS)*, 1292–1295.

Kathuria, S. (2018). E-krishi: An iot based smart irrigation system. *International Journal of Electrical, Electronics and Data Communication*, *6*(6), 8–13.

Kaur, A., Singh, G., Kukreja, V., Sharma, S., Singh, S., & Yoon, B. (2022). Adaptation of IoT with blockchain in Food Supply Chain Management: An analysis-based review in development, benefits and potential applications. *Sensors (Basel), 22*(21), 8174. doi:10.3390/s22218174 PMID:36365871

Kauser, S. H., & Tanguturi, R. C. (2023). Implementation of Machine Learning Approach for Detecting Cardiovascular Diseases. *2023 3rd International Conference on Intelligent Technologies (CONIT)*, 1-6. 10.1109/CONIT59222.2023.10205893

Kavitha, R., Kishore Kumar, R., Pandidurai, M., & Senthil Kamalesh, M. S. C. (2023). Design of IoT based Rural Health Helper using Natural Language Processing. *2023 4th International Conference on Electronics and Sustainable Communication Systems (ICESC)*, 328-333. 10.1109/ICESC57686.2023.10193519

Kawale, S. R., Prasad, K., Palanikkumar, D., Mary, P. A., & Begum, A. Y. (2023). A Novel IoT Framework and Device Architecture for Efficient Smart city Implementation. *2023 7th International Conference on Trends in Electronics and Informatics (ICOEI)*, 420-426. 10.1109/ICOEI56765.2023.10125677

Kawale, S. R., Prasad, K., Anil Kumar, N., Reddy, N. S., & Ashreetha, B. (2023). Technologies for Comprehensive Information Security in the IoT. *2023 International Conference for Advancement in Technology (ICONAT)*, 1-5. 10.1109/ICONAT57137.2023.10080332

Khanna, A., & Anand, R. (2016). IoT based smart parking system. *International Conference on Internet of Things and Applications (IOTA)*, 266-270. doi: 10.1109/IOTA.2016.756273

Khanna, A., Sah, A., Bolshev, V., Burgio, A., Panchenko, V., & Jasiński, M. (2022). Blockchain–Cloud Integration: A Survey. *Sensors (Basel), 22*(14), 5238. doi:10.3390/s22145238 PMID:35890918

Khansalee, E., & Nuanyai, K. (2015). High-Frequency Rectifier for RF Energy Harvesting Systems. *International Conference on Information Technology and Electrical Engineering (ICITEE)*. 10.1109/ICITEED.2015.7408961

Khoshimov, J. (2019). Analysis of the routing methods and Internet models for IoT. In *2019 International Conference on Information Science and Communications Technologies (ICISCT)* (pp. 1-5). IEEE. 10.1109/ICISCT47635.2019.9011968

Khriji, S. (2021). *Precision irrigation: an iot-enabled wireless sensor network for smart irrigation systems. In Women in precision agriculture*. Springer.

Kim, W., Park, J., Yoo, J., Kim, H., & Park, C. (2013). Target localization using ensemble support vector regression in wireless sensor networks. *IEEE Transactions on Cybernetics, 43*(4), 1189–1198. doi:10.1109/TSMCB.2012.2226151 PMID:26502429

Kokkonis, G., Kontogiannis, S., & Tomtsis, D. (2017). A smart IoT fuzzy irrigation system. *Power, 100*(63), 25.

Kumar, A. (2018). A hybrid fuzzy system based cooperative scalable and secured localization scheme for wireless sensor networks. *International Journal of Wireless & Mobile Networks, 10*(3), 51–68. doi:10.5121/ijwmn.2018.10305

Kumar, A., Bhatt, B. R., Anitha, P., Yadav, A. K., Devi, K. K., & Joshi, V. C. (2022). A new diagnosis using a parkinson's disease xgboost and cnn-based classification model using ml techniques. *2022 International Conference on Advanced Computing Technologies and Applications (ICACTA)*, 1–6.

Kumari, P., & Kaur, P. (2021). A survey of fault tolerance in cloud computing. *Journal of King Saud University. Computer and Information Sciences*, *33*(10), 1159–1176. doi:10.1016/j.jksuci.2018.09.021

Kumar, N., Mishra, D., Kim, S. Y., & Jin, S. H. (2020). Self-assembled nis-sns heterostructure via facile successive adsorption and reaction method for high-performance solid-state asymmetric supercapacitors. *Thin Solid Films, 709*, 138138. doi:10.1016/j.tsf.2020.138138

Kumar, P., Dwivedi, R., & Tyagi, V. (2019). Fuzzy Ant Colony Optimization Based Energy Efficient Routing For Mixed Wireless Sensor Network. In *2019 International Conference on Issues and Challenges in Intelligent Computing Techniques (ICICT)* (pp. 1-7). IEEE. 10.1109/ICICT46931.2019.8977699

Kumar, R., Louzir, A., & Naour, J. Y. L. (2016). RFID-based solution for the sensing of home electrical devices activity. *International Symposium on Antenna and Propagation (ISAP2016).*

Lahmiri, S., & Shmuel, A. (2019). Detection of parkinson's disease based on voice patterns ranking and optimized support vector machine. *Biomedical Signal Processing and Control, 49*, 427–433. doi:10.1016/j.bspc.2018.08.029

Laksiri, H. (2019). Design and optimization of iot based smart irrigation system in sri lanka. *14th Conference on Industrial and Information Systems (ICIIS),* 198–202. 10.1109/ICIIS47346.2019.9063272

Laugier, Fraichard, Garnier, & Paromtchik. (1999). *Sensor-based control architecture for a car-like vehicle*. Springer.

Leh, N. A. M. (2019). Smart irrigation system using internet of things. *9th International Conference on System Engineering and Technology (ICSET),* 96–101. 10.1109/ICSEngT.2019.8906497

Leu, J.-S., Chiang, T.-H., Yu, M.-C., & Su, K.-W. (2015). Feburary). Energy Efficient Clustering Scheme for Prolonging the Lifetime of Wireless Sensor Network With Isolated Nodes. *IEEE Communications Letters, 19*(2), 259–262. doi:10.1109/LCOMM.2014.2379715

Liu, L., Gudmundsson, L., Hauser, M., Qin, D., Li, S., & Seneviratne, S. I. (2020). Soil moisture dominates dryness stress on ecosystem production globally. *Nature Communications, 11*(1), 4892. doi:10.1038/s41467-020-18631-1 PMID:32994398

Liu, S., Li, L., Gao, W., Zhang, Y., Liu, Y., Wang, S., & Lu, J. (2018). Diagnosis of Nitrogen Status in Winter Oilseed Rape (Brassica Napus L.) Using In-situ Hyperspectral Data and Unmanned Aerial Vehicle (UAV) Multispectral Images. *Computers and Electronics in Agriculture, 151*, 185–195. doi:10.1016/j.compag.2018.05.026

Liu, Y., Lyu, Q., He, S., Yi, S., Liu, X., Xie, R., Zheng, Y., & Deng, L. (2015). Prediction of Nitrogen and Phosphorus Contents in Citrus Leaves Based on Hyperspectral Imaging. *International Journal of Agricultural and Biological Engineering, 8*, 80–88. doi:10.3965/j.ijabe.20150802.1464

Li, W. (2020). *Review of sensor network-based irrigation systems using iot and remote sensing*. Advances in Meteorology.

Li, Y., Chen, D., Walker, C. N., & Angus, J. F. (2010). Estimating the Nitrogen Status of Crops Using a Digital Camera. *Field Crops Research, 118*(3), 221–227. doi:10.1016/j.fcr.2010.05.011

Loh, T. H., & Cheadle, D. (2017). An Electronically Steerable Parasitic Array Radiator Antenna with Adaptive Beamforming Ability (A Radiation Pattern Assessment). *International Workshop on Electronics: Applications and Student Innovation Competition (iWEM).*

Lotf, J. J., Bonab, M. N., & Khorsandi, S. (2008). A Novel cluster-based routing protocol with extending lifetime for wireless sensor networks. In *2008 5th IFIP International Conference on Wireless and Optical Communications Networks (WOCN '08)* (pp. 1-5). IEEE. 10.1109/WOCN.2008.4542499

Lu, N., Wang, W., Zhang, Q., Li, D., Yao, X., Tian, Y., Zhu, Y., Cao, W., Baret, F., Liu, S., & Cheng, T. (2019). Estimation of Nitrogen Nutrition Status in Winter Wheat from Unmanned Aerial Vehicle Based Multi-angular Multispectral Imagery. *Frontiers in Plant Science, 10*, 1601. doi:10.3389/fpls.2019.01601 PMID:31921250

M, E., RS, R., M, H., K, S., & O, A. (2020). Swarm intelligence–based energy efficient clustering with multihop routing protocol for sustainable wireless sensor networks. *International Journal of Distributed Sensor Networks, 16*(9), 1-12. doi:10.1177/1550147720949133

Mabrouki, J., Azrour, M., Dhiba, D., Farhaoui, Y., & Hajjaji, S. E. (2021, March). IoT-based data logger for weather monitoring using arduino-based wireless sensor networks with remote graphical application and alerts. *Big Data Mining and Analytics, 4*(1), 25–32. doi:10.26599/BDMA.2020.9020018

Mahdavinejad, M., Rezvan, M., Barekatain, M., Adibi, P., Barnaghi, P., & Sheth, A. (2017). Machine learning for Internet of Things data analysis: A survey. *Digital Communications and Networks*. Advance online publication. doi:10.1016/j.dcan.2017.10.002

Maheswari, R., Azath, H., Sharmila, P., & Sheeba Rani Gnanamalar, S. (2019). Smart Village: Solar Based Smart Agriculture with IoT Enabled for Climatic Change and Fertilization of Soil. *2019 IEEE 5th Int. Conf. Mechatronics Syst. Robot. ICMSR 2019*, 102–105. 10.1109/ICMSR.2019.8835454

Maimaitijiang, M., Ghulam, A., Sidike, P., Hartling, S., Maimaitiyiming, M., Peterson, K., Shavers, E., Fishman, J., Peterson, J., Kadam, S., Burken, J., & Fritschi, F. (2017). Unmanned Aerial System (Uas)-based Phenotyping of Soybean Using Multi-sensor Data Fusion and Extreme Learning Machine. *ISPRS Journal of Photogrammetry and Remote Sensing*, *134*, 43–58. doi:10.1016/j.isprsjprs.2017.10.011

Makarious, M. B., Leonard, H. L., Vitale, D., Iwaki, H., Sargent, L., Dadu, A., Violich, I., Hutchins, E., Saffo, D., & Bandres-Ciga, S. (2022). Multi- modality machine learning predicting parkinson's disease. *NPJ. Parkinson's Disease*, *8*(1), 35. PMID:35365675

Mali, Tirmare, Suryawanshi, & Chaturvedi. (2022). Novel Predictive Control and Monitoring System based on IoT for Evaluating Industrial Safety Measures. *IJEER, 10*(4), 1050-1057. doi:10.37391/ijeer.100448

Manimegalai, V., Judy, A. L., Gayathri, A., Ashadevi, S., & Mohanapriya, V. (2020). Smart irrigation system with monitoring and controlling using iot. *International Journal of Engineering and Advanced Technology*, *9*(4), 1373–1376. doi:10.35940/ijeat.C6586.049420

Maresma, Á., Lloveras, J., & Martínez-Casasnovas, J. A. (2018). Use of multispectral airborne images to improve in-season nitrogen management, predict grain yield and estimate economic return of maize in irrigated high yielding environments. *Remote Sensing (Basel)*, *10*(4), 543. doi:10.3390/rs10040543

Mary, D. L., & Ramakrishnan, M. (2021). A Novel Approach to Optimize Water and Fertilizers in Agriculture using IoT. *Int. J. Cybern. Informatics*, *10*(2), 57–64. doi:10.5121/ijci.2021.100207

Mei, J., Desrosiers, C., & Frasnelli, J. (2021). Machine learning for the diagnosis of parkinson's disease: A review of literature. *Frontiers in Aging Neuroscience*, *13*, 633752. doi:10.3389/fnagi.2021.633752 PMID:34025389

Mercado-Luna, A., Rico-García, E., Lara-Herrera, A., Soto-Zarazúa, G., Ocampo-Velázquez, R., Guevara-González, R., Herrera-Ruiz, G., & Torres-Pacheco, I. (2010). Nitrogen Determination on Tomato (Lycopersicon Esculentum Mill.) Seedlings by Color Image Analysis (RGB). *African Journal of Biotechnology*, *9*, 5326–5332.

MINAGRI. (2020). *Minagri Annu. Rep. 2019-2020.* Author.

Mirjalili, S. (2016). SCA: A Sine Cosine Algorithm for solving optimization problems. *Knowledge-Based Systems*, *96*, 120–133. doi:10.1016/j.knosys.2015.12.022

Mishra, D., Abbas, A., Pande, T., Pandey, A. K., Agrawal, K. K., & Yadav, R. S. (2019). Smart agriculture system using IoT. *ACM Int. Conf. Proceeding Ser.*

Mittal, V., & Sharma, R. (2021). Classification of parkinson disease based on analysis and synthesis of voice signal. *International Journal of Healthcare Information Systems and Informatics*, *16*(4), 1–22. doi:10.4018/IJHISI.20211001.oa30

Mody, S., Mirkar, S., Ghag, R., & Kotecha, P. (2021). Cluster Head Selection Algorithm For Wireless Sensor Networks Using Machine Learning. In *2021 International Conference on Computational Performance Evaluation (ComPE)* (pp. 445-450). IEEE. 10.1109/ComPE53109.2021.9752264

Moharana & Dutta. (2016). Spatial Variability of Chlorophyll and Nitrogen Content of Rice from Hyperspectral Imagery. *ISPRS J. Photogramm. Remote Sens, 122*, 17–29. doi:.isprsjprs.2016.09.002 doi:10.1016/j

Mottaghinia, Z., & Ghaffari, A. (2018). *Fuzzy logic based distance and energy-aware routing protocol in delay-tolerant mobile sensor networks*. Academic Press.

Mousa, A. K., Croock, M. S., & Abdullah, M. N. (2014). Fuzzy based decision support model for irrigation system management. *International Journal of Computer Applications*, *104*(9). Advance online publication. doi:10.5120/18230-9177

Mudholkar, P., Mudholkar, M., Puneeth Kumar, B. S., Dankan Gowda, V., & Raju, S. S. (2021). Smart Villages: IoT Technology Based Transformation. *Journal of Physics: Conference Series*, *2070*(1), 012128. doi:10.1088/1742-6596/2070/1/012128

Multilevel Vehicle Allotment And Retrieval System Using IoT. (2022). *International Research Journal of Modernization in Engineering Technology and Science* , 4(3).

Muneeswari. (2017). Smart irrigation system using iot approach. *International Journal of Engineering Research & Science*.

Munir, M. S. (2021). *Intelligent and smart irrigation system using edge computing and iot*. Complexity.

N., Farooq, M, Reddy, & Tanguturi. (2022). Improved Secure Communication Mechanism for IoT Platform Devices. *2022 International Conference on Smart Generation Computing, Communication and Networking (SMART GENCON)*, 1-6. 10.1109/SMARTGENCON56628.2022.10083714

Naik, P. (2018). Automation of irrigation system using iot. International Journal of Engineering and Manufacturing Science, 8(1), 77–88.

Naveena Pai, Swathi Pai, Shruthi, & Naveen. (2020). Internet of Things: A Survey on Devices, Ecosystem, Components and Communication Protocols. *2020 4th International Conference on Electronics, Communication and Aerospace Technology (ICECA)*, 611-616. . doi:10.1109/ICECA49313.2020.9297458

Nguyen, T., Vo, K. T., Ta, T. T., Nguven-Hoang, T. A., & Dinh, N. T. (2022, October). The benefits and challenges of applying Blockchain technology into Big Data: A literature review. In *2022 International Conference on Advanced Technologies for Communications (ATC)* (pp. 219-224). IEEE. 10.1109/ATC55345.2022.9943004

Noor, M. Z. M., & Ramlee, R. A. (2021). Performances Analysis of IoT Based Smart Greenhouse System. *Int. J. Electr. Eng. Appl. Sci.*, *4*(2), 1–8.

Noor, N. M., Razak, Z., & Yamani, M. (2009). Car Parking System: A Review of Smart Parking System and its Technology|| -. *Information Technology Journal.*

Orn, D., Duan, L., Liang, Y., Siy, H., & Subramaniam, M. (2020). Agro-AI Education: Artificial Intelligence for Future Farmers. *SIGITE 2020 - Proc. 21st Annu. Conf. Inf. Technol. Educ.*, 54–57. 10.1145/3368308.3415457

Osco, L. P., Ramos, A. P. M., Pereira, D. R., Éas, M., Imai, N. N., Matsubara, E. T., & Estrabis, N. (2019). Predicting Canopy Nitrogen Content in Citrus-trees Using Random Forest Algorithm Associated to Spectral Vegetation Indices from UAV-Imagery. *Remote Sensing (Basel), 11*(24), 2925. doi:10.3390/rs11242925

Padalalu, P., Mahajan, S., Dabir, K., Mitkar, S., & Javale, D. (2017). Smart water dripping system for agriculture/farming. *2017 2nd International Conference for Convergence in Technology (I2CT),* 659–662. 10.1109/I2CT.2017.8226212

Palconit, M. G. B. (2020). Iot-based precision irrigation system for eggplant and tomato. *9th Int. Symp. Comput. Intell. Ind. Appl.*

Panchal, A., Singh, L., & Singh, R. K. (2020). RCH-LEACH: Residual Energy based Cluster Head Selection in LEACH for Wireless Sensor Networks. In *2020 International Conference on Electrical and Electronics Engineering (ICE3)* (pp. 322-325). IEEE. 10.1109/ICE348803.2020.9122962

Pandey, P., Ge, Y., Stoerger, V., & Schnable, J. C. (2017). High Throughput in Vivo Analysis of Plant Leaf Chemical Properties Using Hyperspectral Imaging. *Frontiers in Plant Science, 8,* 1348. doi:10.3389/fpls.2017.01348 PMID:28824683

Patan, R., Parizi, R. M., Dorodchi, M., Pouriyeh, S., & Rorrer, A. (2023). *Blockchain Education: Current State, Limitations, Career Scope, Challenges, and Future Directions.* arXiv preprint arXiv:2301.07889.

Pathak, A., AmazUddin, M., Abedin, M. J., Andersson, K., Mustafa, R., & Hossain, M. S. (2019). Iot based smart system to support agricultural parameters: A case study. *Procedia Computer Science, 155,* 648–653. doi:10.1016/j.procs.2019.08.092

Patidar, Y., Jain, M., & Vyas, A. K. (2022a). Comparative Analysis of Clustering and Routing Protocols for Energy-Efficient WSN. In *2022 4th International Conference on Advances in Computing, Communication Control and Networking (ICAC3N)* (pp. 1768-1773). IEEE. 10.1109/ICAC3N56670.2022.10074074

Patidar, Y., Jain, M., & Vyas, A. K. (2022b). Modified Unexpended Energy Based Stable Cluster Head Selection for Homogeneous Wireless Sensor Networks (WSNs). In *2022 3rd International Conference on Computing, Analytics and Networks (ICAN)* (pp. 1-5). IEEE. 10.1109/ICAN56228.2022.10007098

Perea, R. G., Poyato, E. C., Montesinos, P., & Díaz, J. A. R. (2019). Prediction of irrigation event occurrence at farm level using optimal decision trees. *Computers and Electronics in Agriculture.*

Perinbam. (n.d.). Automatic Irrigation System using Fuzzy Logic. *Hemosphere, 89*(7), 884-892.

Phoemphon, S., So-In, C., & Niyato, D. (2018). A hybrid model using fuzzy logic and an extreme learning machine with vector particle swarm optimization for wireless sensor network localization. *Applied Soft Computing*, *65*, 101–120. doi:10.1016/j.asoc.2018.01.004

Podder. (2021). Iot based smart agrotech system for verification of urban farming parameters. *Microprocessors and Microsystems, 82.*

Poirier, A.-A., Aub'e, B., C^ot'e, M., Morin, N., Di Paolo, T., & Soulet, D. (2016). Gastrointestinal dysfunctions in parkinson's disease: Symptoms and treatments. *Parkinson's Disease*, *2016*, 2016. doi:10.1155/2016/6762528 PMID:28050310

Polat, K. (2019). A hybrid approach to parkinson disease classification using speech signal: the combination of smote and random forests. 2019 scientific meeting on electrical-electronics & biomedical engineering and computer science (EBBT), 1–3. doi:10.1109/EBBT.2019.8741725

Prabha, R., Sinitambirivoutin, E., Passelaigue, F., & Ramesh, M. V. (2018). Design and Development of an IoT Based Smart Irrigation and Fertilization System for Chilli Farming. *2018 Int. Conf. Wirel. Commun. Signal Process. Networking, WiSPNET 2018*, 1–7. 10.1109/WiSPNET.2018.8538568

Pradeep, K., & Balasundaram, A. (2021). Iot based smart irrigation for agricultural fields. *Annals of the Romanian Society for Cell Biology*, 2000–2009.

Prakash, B. R. (2020). Super smart irrigation system using internet of things. *2020 7th International Conference on Smart Structures and Systems (ICSSS),* 1–5. 10.1109/ICSSS49621.2020.9202361

Qamar, R., & Zardari, B. A. (2023). A Study of Blockchain-Based Internet of Things. *Iraqi Journal For Computer Science and Mathematics*, *4*(1), 15–23.

R. M., Nareshkumar, & Gururaj. (2023). An Integrated IoT Technology for Health and Traffic Monitoring System with Smart Ambulance. *2022 IEEE North Karnataka Subsection Flagship International Conference (NKCon)*, 1-6. 10.1109/NKCon56289.2022.10126786

Rabbani, M. R., Hassan, M. K., Hudaefi, F. A., & Shaikh, Z. H. (2022). Islamic finance and cryptocurrency: a systematic review. *FinTech in Islamic Financial Institutions: Scope, Challenges, and Implications in Islamic Finance,* 279-306.

Rajkumar, M. N. (2017). Intelligent irrigation systema^C"an iot based approach. *International Conference on Innovations in Green Energy and Healthcare Technologies (IGEHT),* 1–5.

Ramesha, M., Jeevan, K., & Sathisha, B. M. (2020). Implementation of IoT Based Wireless Electronic Stethoscope. *2020 Third International Conference on Multimedia Processing, Communication & Information Technology (MPCIT)*, 103-106. 10.1109/MPCIT51588.2020.9350476

Ramya, R., & Ramamoorthy, S. (2022). Analysis of machine learning algorithms for efficient cloud and edge computing in the loT. Challenges and Risks Involved in Deploying 6G and NextGen Networks, 72–90.

Ramya, R., & Ramamoorthy, S. (2022). Development of a framework for adaptive productivity management for edge computing based IoT applications. *AIP Conference Proceedings, 2519*, 030068. doi:10.1063/5.0111710

Ramya, R., & Ramamoorthy, S. (2022). Survey on Edge Intelligence in IoT-Based Computing Platform. *Lecture Notes in Networks and Systems, 356*, 549–561. doi:10.1007/978-981-16-7952-0_52

Ramya, R., & Ramamoorthy, S. (2023). Lightweight Unified Collaborated Relinquish Edge Intelligent Gateway Architecture with Joint Optimization. *IEEE Access : Practical Innovations, Open Solutions*, *11*, 90396–90409. doi:10.1109/ACCESS.2023.3307808

Ramya, R., & Ramamoorthy, S. (2023). QoS in multimedia application for IoT devices through edge intelligence. *Multimedia Tools and Applications*. Advance online publication. doi:10.1007/s11042-023-15941-6

Rana, A., Dumka, A., Singh, R., Panda, M. K., Priyadarshi, N., & Twala, B. (2022). Imperative role of machine learning algorithm for detection of parkinson's disease: Review, challenges and recommendations. *Diagnostics (Basel)*, *12*(8), 2003. doi:10.3390/diagnostics12082003 PMID:36010353

Rawal, S. (2017). Iot based smart irrigation system. *International Journal of Computer Applications*, *159*(8), 7–11. doi:10.5120/ijca2017913001

Rehman, A., Saba, T., Kashif, M., Fati, S. M., Bahaj, S. A., & Chaudhry, H. (2022). A Revisit of Internet of Things Technologies for Monitoring and Control Strategies in Smart Agriculture. *Agronomy (Basel)*, *12*(1), 1–21. doi:10.3390/agronomy12010127

Research Nester. (2016). *Internet of things (IoT) Market: Global Demand, Growth Analysis & Opportunity Outlook 2023*. Retrieved from https://www.researchnester.com/reports/internet-of-things-iot-market-global-demand-growth-analysis-opportunity-outlook-2023/216

Rizvi, D. R., Nissar, I., Masood, S., Ahmed, M., & Ahmad, F. (2020). An lstm based deep learning model for voice-based detection of parkinson's disease. *Int. J. Adv. Sci. Technol*, *29*(8).

S, E. F., T C, E. S., & R S, V. D. (2022). Prediction of parkinson's disease using xgboost. *2022 8th International Conference on Advanced Computing and Communication Systems (ICACCS)*, *1*, 1769–1772.

Sachin, P. (2019). Iot based smart irrigation system. *International Journal of Advanced Research in Electrical, Electronics and Instrumentation Engineering*, *8*(5), 1604–1608.

Salama, R., & Al-Turjman, F. (2022, August). AI in Blockchain Towards Realizing Cyber Security. In *2022 International Conference on Artificial Intelligence in Everything (AIE)* (pp. 471-475). IEEE. 10.1109/AIE57029.2022.00096

Saranya, V., Shankar, S., & Kanagachidambaresan, G. R. (2018). Energy Efficient Clustering Scheme (EECS) for Wireless Sensor Network with Mobile Sink. *Wireless Personal Communications, 100*(4), 1553–1567. doi:10.1007/s11277-018-5653-1

Sathiya, V., Nagalakshmi, K., Jeevamalar, J., Babu, R. A., Karthi, R., Acevedo-Duque, Á., ... Ramabalan, S. (2023). Reshaping healthcare supply chain using chain-of-things technology and key lessons experienced from COVID-19 pandemic. *Socio-Economic Planning Sciences, 85*, 101510. doi:10.1016/j.seps.2023.101510 PMID:36687377

Satyanarayana, S. K., Akhil, A., & Padmini, G. (2021). Smart parking assistance using Arduino. *International Journal for Research in Applied Science & Engineering Technology, 9*(6).

Sayed, M. A. (2019). An iot based robotic system for irrigation notifier. *International Conference on Robotics, Automation, Artificial-intelligence and Internet-of-Things (RAAICON),* 77–80. 10.1109/RAAICON48939.2019.38

Segal, M. R. (2004). *Machine learning benchmarks and random forest regression*. Academic Press.

Seignez, E., Lambert, A., & Maurin, T. (2005). Autonomous parking carrier for intelligent vehicle. *IEEE Proceedings. Intelligent Vehicles Symposium*, 411-416. doi: 10.1109/IVS.2005.1505138

Shabbir, N., & Hassan, S. R. (2017). Routing Protocols for Wireless Sensor Networks (WSNs). In Wireless Sensor Networks - Insights and Innovations. IntechOpen. doi:10.5772/intechopen.70208

Shah, K., Patel, N., Thakkar, J., & Patel, C. (2022). Exploring applications of blockchain technology for Industry 4.0. *Materials Today: Proceedings, 62*, 7238–7242. doi:10.1016/j.matpr.2022.03.681

Sharawi, M., & Emary, E. (2017). Impact of grey wolf optimization on WSN cluster formation and lifetime expansion. In *2017 Ninth International Conference on Advanced Computational Intelligence (ICACI)* (pp. 157-162). IEEE. 10.1109/ICACI.2017.7974501

Sharma, P. A., Reddy, S., & Patwal, P. P. S. (2022). Data Analytics and Cloud-Based Platform for Internet of Things Applications in Smart Cities. *2022 International Conference on Industry 4.0 Technology (I4Tech)*, 1-6. 10.1109/I4Tech55392.2022.9952780

Shendryk, Y., Sofonia, J., Garrard, R., Rist, Y., Skocaj, D., & Thorburn, P. (2020). Fine-scale Prediction of Biomass and Leaf Nitrogen Content in Sugarcane Using UAV LiDAR and Multispectral Imaging. *International Journal of Applied Earth Observation and Geoinformation, 92*, 102177. doi:10.1016/j.jag.2020.102177

Sherstinsky, A. (2020). Fundamentals of recurrent neural network (rnn) and long short-term memory (lstm) network. *Physica D. Nonlinear Phenomena, 404*, 132306.

Shufian, A. (2019). Smart irrigation system with solar power and gsm technology. *5th International Conference on Advances in Electrical Engineering (ICAEE),* 301–305. 10.1109/ICAEE48663.2019.8975634

Singh, K. (2019). Iot based approach for smart irrigation system suited to multiple crop cultivation. *International Journal of Engineering Research & Technology (Ahmedabad), 12*, 357–363.

Singh, K. (2021). *Design of a low-cost sensor-based iot system for smart irrigation. In Applications in Ubiquitous Computing*. Springer.

Soni, S., & Shrivastava, M. (2018). Novel learning algorithms for efficient mobile sink data collection using reinforcement learning in wireless sensor network. Wireless Communications and Mobile Computing. doi:10.1155/2018/7560167

Soumaya, Z., Taoufiq, B. D., Benayad, N., Yunus, K., & Abdelkrim, A. (2021). The detection of parkinson disease using the genetic algorithm and svm classifier. *Applied Acoustics*, *171*, 107528. doi:10.1016/j.apacoust.2020.107528

Sridhara, S. B., Naveen, K. B., Ramesha, M., & Pai, G. N. (2020). Internet of things: Internet revolution, impact, technology road map and features. *Adv. Math. Sci. J.*, *9*(7), 4405–4414. doi:10.37418/amsj.9.7.11

Statista. (2018). *Internet of Things (IoT) connected devices installed base worldwide from 2015 to 2025 (in billions)*. Retrieved from https://www.statista.com/statistics/471264/iot-number-of-connected-devices-worldwide/

Subramani, C. (2020). *Iot-based smart irrigation system. In Cognitive Informatics and Soft Computing*. Springer.

Sudharshan, N., Karthik, A. V. S. K., Kiran, J. S. S., & Geetha, S. (2019). Renewable energy based smart irrigation system. *Procedia Computer Science*, *165*, 615–623. doi:10.1016/j.procs.2020.01.055

Suhaimi, A. F., Yaakob, N., Saad, S. A., & Azami, K. (2021). *IoT Based Smart Agriculture Monitoring*. Automation and Intrusion Detection System IoT Based Smart Agriculture Monitoring, Automation and Intrusion Detection System.

Sun, W., Song, Y., & Chen, M. (2010). A load-balanced and energy-aware routing metric for wireless multimedia sensor networks. In *IET 3rd International Conference on Wireless, Mobile and Multimedia Networks (ICWMNN 2010)* (pp. 21-24). IET. 10.1049/cp.2010.0608

Sun, Y., Zhang, X., & Wang, X. (2018). Device-free wireless localization using artificial neural networks in wireless sensor networks. Wireless Communications and Mobile Computing. doi:10.1155/2018/4201367

Sun, C., Hirata, A., Ohira, T., & Karmakar, N. C. (2004). Fast Beamforming of Electrically Steerable Parasitic Array Radiator Antennas: Theory and Experiment. *IEEE Transactions on Antennas and Propagation*, *52*(7), 1819–1832. doi:10.1109/TAP.2004.831314

Supriya, B. (2021). Iot based remote smart irrigation system. *International Journal of Modern Agriculture*, *10*(2), 3805–3811.

Suvarna, S. (2017). Smart Car Parking System using Arduino UNO. *International Journal of Computer Applications*, *169*(1), 13-18. 10.5120/ijca2017914425

Suvra & Sen. (n.d.). *Iot based automated irrigation and smart agriculture monitoring system.* Academic Press.

Swaminathan, Palani, Kotecha, Kumar, & V. (2022). IoT Driven Artificial Intelligence Technique for Fertilizer Recommendation Model. *IEEE Consum. Electron. Mag.*

Swetha, Namitha, Manu, Rashmi, & Chinamuttevi. (2022). IOT Based Smart Health Care System to Monitor Covid-19 Patients. *IJEER, 10*(1), 36-40. doi:10.37391/IJEER.100105

Syaza, N. B. I. (2008). Smart Home Garden Irrigation System With Raspberry Pi. *IEEE, 16*(June), 24.

Thakur, D., Kumar, Y., & Vijendra, S. (2020). Smart irrigation and intrusions detection in agricultural fields using iot. *Procedia Computer Science, 167*, 154–162. doi:10.1016/j.procs.2020.03.193

Triantafyllou, Sarigiannidis, & Bibi. (2019). Precision agriculture: A remote sensing monitoring system architecture. *Inf., 10*(11).

Tsai, T.-H., Huang, C.-C., Chang, C.-H., & Hussain, M. A. (2020). Design of Wireless Vision Sensor Network for Smart Home. *IEEE Access : Practical Innovations, Open Solutions, 8*, 60455–60467. doi:10.1109/ACCESS.2020.2982438

TsanasA.LittleM.McSharryP.RamigL. (2009). Accurate telemoni- toring of parkinson's disease progression by non-invasive speech tests. Nature Precedings, 1–1.

Ullah, R. (2021). *Eewmp: an iot-based energy-efficient water management platform for smart irrigation.* Scientific Programming.

Ullo, S. L., & Sinha, G. R. (2020). Advances in smart environment monitoring systems using iot and sensors. *Sensors (Basel), 20*(11), 1–18. doi:10.3390/s20113113 PMID:32486411

Vaishali, S., Suraj, S., Vignesh, G., Dhivya, S., & Udhayakumar, S. (2018). Mobile integrated smart irrigation management and monitoring system using IOT. *Proc. 2017 IEEE Int. Conf. Commun. Signal Process. ICCSP 2017,* 2164–2167.

Valdes-Garcia, A., Sadhu, B., Gu, X., Tousi, Y., Liu, D., Scott, K., Hillin, J., Sahl, S., & Rexberg, L. (2018). Circuit and Antenna-in-Package Innovations for Scaled mmWave 5G Phased Array modules. *IEEE Custom Integrated Circuits Conference (CICC).* 10.1109/CICC.2018.8357050

Vellaichamy, J., Basheer, S., Bai, P., Khan, M., Kumar Mathivanan, S., Jayagopal, P., & Babu, J. (2023). Wireless Sensor Networks Based on Multi-Criteria Clustering and Optimal Bio-Inspired Algorithm for Energy-Efficient Routing. *Applied Sciences (Basel, Switzerland), 13*(5), 1–14. doi:10.3390/app13052801

Velmurugan. (2020). *An iot based smart irrigation system using soil moisture and weather prediction.* Academic Press.

Waghmare, P., Chandgude, M., Chaure, P., & Chaudhari, A. (2017). Survey on: Home Automation Systems. *International Conference on Trends in Electronics and Informatics (ICEI2017).*

Wang, B. (2021). Wireless Sensor Network Routing Algorithm Based on Traffic Prediction for Internet of Thing. In *2021 International Conference on Electronic Communications, Internet of Things and Big Data (ICEIB)* (pp. 82-86). IEEE. 10.1109/ICEIB53692.2021.9686397

Wang, D., Cao, W., Zhang, F., Li, Z., Xu, S., & Wu, X. (2022). A Review of Deep Learning in Multiscale Agricultural Sensing. *Remote Sensing (Basel)*, *14*(3), 559. doi:10.3390/rs14030559

Wang, J., Gao, Y., Liu, W., Sangaiah, A., & Kim, H. (2019). An improved routing schema with special clustering using PSO algorithm for heterogeneous wireless sensor network. *Sensors (Basel)*, *19*(3), 671. doi:10.3390/s19030671 PMID:30736392

Wang, L., Wang, J., & Shi, J. (2018). Design of 4x4 Meandering-Fed Stacked Patch Antenna Array. *International Workshop on Antenna Technology (iWAT).*

Xie, T., Huang, Z., Chi, Z., & Zhu, T. (2017). Minimizing amortized cost of the on-demand irrigation system in smart farms. *3rd International Workshop on Cyber-Physical Systems for Smart Water Networks*, 43–46. 10.1145/3055366.3055370

Xu, F., Ye, H., Yang, F., & Zhao, C. (2019, January 04). Software Defined Mission-Critical Wireless Sensor Network: Architecture and Edge Offloading Strategy. *IEEE Access : Practical Innovations, Open Solutions*, *7*, 10383–10391. doi:10.1109/ACCESS.2019.2890854

Xu, S., & Pan, Z. (2020). A novel ensemble of random forest for assisting diagnosis of parkinson's disease on small handwritten dynamics dataset. *International Journal of Medical Informatics*, *144*, 104283. doi:10.1016/j.ijmedinf.2020.104283 PMID:33010729

Yadav, A., Kumar, S., & Vijendra, S. (2018). Network life time analysis of WSNs using particle swarm optimization. *Elsevier*, *132*, 805–815.

Yaga, D., Mell, P., Roby, N., & Scarfone, K. (2019). *Blockchain technology overview*. arXiv preprint arXiv:1906.11078.

Yaman, O., Ertam, F., & Tuncer, T. (2020). Automated parkinson's disease recognition based on statistical pooling method using acoustic features. *Medical Hypotheses*, *135*, 109483. doi:10.1016/j.mehy.2019.109483 PMID:31954340

Ye, X., Abe, S., & Zhang, S. (2020). Estimation and Mapping of Nitrogen Content in Apple Trees at Leaf and Canopy Levels Using Hyperspectral Imaging. *Precision Agriculture*, *21*(1), 198–225. doi:10.1007/s11119-019-09661-x

Younis, O., & Fahmy, S. (2004, October-December). HEED: A hybrid, energy-efficient, distributed clustering approach for ad hoc sensor networks. *IEEE Transactions on Mobile Computing*, *3*(5), 366–379. doi:10.1109/TMC.2004.41

Yousif, Y. K., Badlishah, R., Yaakob, N., & Amir, A. (2017). An Energy Efficient and Load Balancing Clustering Scheme for Wireless Sensor Network (WSN) Based on Distributed Approach. In *J. Phys.: Conf. Ser. 1st International Conference on Green and Sustainable Computing (ICoGeS)* (pp. 1-6). IOP Publishing Ltd. 10.1088/1742-6596/1019/1/012007

Yu, X., Lu, H., & Liu, Q. (2018). Deep-learning-based Regression Model and Hyperspectral Imaging for Rapid Detection of Nitrogen Concentration in Oilseed Rape (Brassica Napus L.) Leaf. *Chemometrics and Intelligent Laboratory Systems*, *172*, 188–193. doi:10.1016/j.chemolab.2017.12.010

Zagrouba, R., & Kardi, A. (2021). Comparative Study of Energy Efficient Routing Techniques in Wireless Sensor Networks. *Information (Basel)*, *12*(1), 1–28. doi:10.3390/info12010042

Zang, Z., Qi, J.-D., & Cao, Y.-J. (2010). A Robust Routing Protocol in wireless sensor network. In *IET International Conference on Wireless Sensor Network 2010 (IET-WSN 2010)* (pp. 276-279). IET. 10.1049/cp.2010.1066

Zhang, D., Song, H., Cheng, H., Hao, D., Wang, H., Kan, G., Jin, H., & Yu, D. (2014). The Acid Phosphatase-encoding Gene GmACP1 Contributes to Soybean Tolerance to Low-phosphorus Stress. *PLOS Genetics*, *10*(1), 1. doi:10.1371/journal.pgen.1004061 PMID:24391523

Zhang, X., Tan, Y., Ling, Y., Lu, G., Liu, F., Yi, Z., Jia, X., Wu, M., Shi, B., Xu, S., Chen, J., Wang, W., Chen, B., Jiang, L., Yu, S., Lu, J., Wang, J., Xu, M., Yuan, Z., ... Lu, H. (2020). Viral and host factors related to the clinical outcome of COVID-19. *Nature*, *583*(7816), 437–440. doi:10.1038/s41586-020-2355-0 PMID:32434211

Zhao, W. (2017). *Design and implementation of smart irrigation system based on lora. In IEEE Globecom Workshops (GC Wkshps)*. IEEE.

Zheng, H., Cheng, T., Li, D., Yao, X., Tian, Y., Cao, W., & Zhu, Y. (2018). Combining Unmanned Aerial Vehicle (Uav)-based Multispectral Imagery and Ground-based Hyperspectral Data for Plant Nitrogen Concentration Estimation in Rice. *Frontiers in Plant Science*, *9*, 936. doi:10.3389/fpls.2018.00936 PMID:30034405

Zhixiang, D., & Bensheng, Q. (2007). Three-layered routing protocol for WSN based on LEACH algorithm. In *IET Conference on Wireless, Mobile and Sensor Networks 2007 (CCWMSN07)* (pp. 72-75). IET. 10.1049/cp:20070086

Zodpe, H., & Sapkal, A. (2020). An efficient AES implementation using FPGA with enhanced security features. *Journal of King Saud University. Engineering Sciences*, *32*(2), 115–122. doi:10.1016/j.jksues.2018.07.002

Zulfiqar, U., Jiang, W., Xiukang, W., Hussain, S., Ahmad, M., Maqsood, M. F., Ali, N., Ishfaq, M., Kaleem, M., Haider, F. U., Farooq, N., Naveed, M., Kucerik, J., Brtnicky, M., & Mustafa, A. (2022). Cadmium phytotoxicity, tolerance, and advanced remediation approaches in agricultural soils; a comprehensive review. *Frontiers in Plant Science*, *13*, 773815. doi:10.3389/fpls.2022.773815 PMID:35371142

Related References

To continue our tradition of advancing information science and technology research, we have compiled a list of recommended IGI Global readings. These references will provide additional information and guidance to further enrich your knowledge and assist you with your own research and future publications.

Aasi, P., Rusu, L., & Vieru, D. (2017). The Role of Culture in IT Governance Five Focus Areas: A Literature Review. *International Journal of IT/Business Alignment and Governance, 8*(2), 42-61. https://doi.org/ doi:10.4018/IJITBAG.2017070103

Abdrabo, A. A. (2018). Egypt's Knowledge-Based Development: Opportunities, Challenges, and Future Possibilities. In A. Alraouf (Ed.), *Knowledge-Based Urban Development in the Middle East* (pp. 80–101). Hershey, PA: IGI Global. doi:10.4018/978-1-5225-3734-2.ch005

Abu Doush, I., & Alhami, I. (2018). Evaluating the Accessibility of Computer Laboratories, Libraries, and Websites in Jordanian Universities and Colleges. *International Journal of Information Systems and Social Change, 9*(2), 44–60. doi:10.4018/IJISSC.2018040104

Adegbore, A. M., Quadri, M. O., & Oyewo, O. R. (2018). A Theoretical Approach to the Adoption of Electronic Resource Management Systems (ERMS) in Nigerian University Libraries. In A. Tella & T. Kwanya (Eds.), *Handbook of Research on Managing Intellectual Property in Digital Libraries* (pp. 292–311). Hershey, PA: IGI Global. doi:10.4018/978-1-5225-3093-0.ch015

Afolabi, O. A. (2018). Myths and Challenges of Building an Effective Digital Library in Developing Nations: An African Perspective. In A. Tella & T. Kwanya (Eds.), *Handbook of Research on Managing Intellectual Property in Digital Libraries* (pp. 51–79). Hershey, PA: IGI Global. doi:10.4018/978-1-5225-3093-0.ch004

Agarwal, P., Kurian, R., & Gupta, R. K. (2022). Additive Manufacturing Feature Taxonomy and Placement of Parts in AM Enclosure. In S. Salunkhe, H. Hussein, & J. Davim (Eds.), *Applications of Artificial Intelligence in Additive Manufacturing* (pp. 138–176). IGI Global. https://doi.org/10.4018/978-1-7998-8516-0.ch007

Al-Alawi, A. I., Al-Hammam, A. H., Al-Alawi, S. S., & AlAlawi, E. I. (2021). The Adoption of E-Wallets: Current Trends and Future Outlook. In Y. Albastaki, A. Razzaque, & A. Sarea (Eds.), *Innovative Strategies for Implementing FinTech in Banking* (pp. 242–262). IGI Global. https://doi.org/10.4018/978-1-7998-3257-7.ch015

Alsharo, M. (2017). Attitudes Towards Cloud Computing Adoption in Emerging Economies. *International Journal of Cloud Applications and Computing*, *7*(3), 44–58. doi:10.4018/IJCAC.2017070102

Amer, T. S., & Johnson, T. L. (2017). Information Technology Progress Indicators: Research Employing Psychological Frameworks. In A. Mesquita (Ed.), *Research Paradigms and Contemporary Perspectives on Human-Technology Interaction* (pp. 168–186). Hershey, PA: IGI Global. doi:10.4018/978-1-5225-1868-6.ch008

Andreeva, A., & Yolova, G. (2021). Liability in Labor Legislation: New Challenges Related to the Use of Artificial Intelligence. In B. Vassileva & M. Zwilling (Eds.), *Responsible AI and Ethical Issues for Businesses and Governments* (pp. 214–232). IGI Global. https://doi.org/10.4018/978-1-7998-4285-9.ch012

Anohah, E. (2017). Paradigm and Architecture of Computing Augmented Learning Management System for Computer Science Education. *International Journal of Online Pedagogy and Course Design*, *7*(2), 60–70. doi:10.4018/IJOPCD.2017040105

Anohah, E., & Suhonen, J. (2017). Trends of Mobile Learning in Computing Education from 2006 to 2014: A Systematic Review of Research Publications. *International Journal of Mobile and Blended Learning*, *9*(1), 16–33. doi:10.4018/IJMBL.2017010102

Arbaiza, C. S., Huerta, H. V., & Rodriguez, C. R. (2021). Contributions to the Technological Adoption Model for the Peruvian Agro-Export Sector. *International Journal of E-Adoption, 13*(1), 1–17. https://doi.org/10.4018/IJEA.2021010101

Bailey, E. K. (2017). Applying Learning Theories to Computer Technology Supported Instruction. In M. Grassetti & S. Brookby (Eds.), *Advancing Next-Generation Teacher Education through Digital Tools and Applications* (pp. 61–81). Hershey, PA: IGI Global. doi:10.4018/978-1-5225-0965-3.ch004

Baker, J. D. (2021). Introduction to Machine Learning as a New Methodological Framework for Performance Assessment. In M. Bocarnea, B. Winston, & D. Dean (Eds.), *Handbook of Research on Advancements in Organizational Data Collection and Measurements: Strategies for Addressing Attitudes, Beliefs, and Behaviors* (pp. 326–342). IGI Global. https://doi.org/10.4018/978-1-7998-7665-6.ch021

Banerjee, S., Sing, T. Y., Chowdhury, A. R., & Anwar, H. (2018). Let's Go Green: Towards a Taxonomy of Green Computing Enablers for Business Sustainability. In M. Khosrow-Pour (Ed.), *Green Computing Strategies for Competitive Advantage and Business Sustainability* (pp. 89–109). Hershey, PA: IGI Global. doi:10.4018/978-1-5225-5017-4.ch005

Basham, R. (2018). Information Science and Technology in Crisis Response and Management. In M. Khosrow-Pour, D.B.A. (Ed.), Encyclopedia of Information Science and Technology, Fourth Edition (pp. 1407-1418). Hershey, PA: IGI Global. doi:10.4018/978-1-5225-2255-3.ch121

Batyashe, T., & Iyamu, T. (2018). Architectural Framework for the Implementation of Information Technology Governance in Organisations. In M. Khosrow-Pour, D.B.A. (Ed.), Encyclopedia of Information Science and Technology, Fourth Edition (pp. 810-819). Hershey, PA: IGI Global. doi:10.4018/978-1-5225-2255-3.ch070

Bekleyen, N., & Çelik, S. (2017). Attitudes of Adult EFL Learners towards Preparing for a Language Test via CALL. In D. Tafazoli & M. Romero (Eds.), *Multiculturalism and Technology-Enhanced Language Learning* (pp. 214–229). Hershey, PA: IGI Global. doi:10.4018/978-1-5225-1882-2.ch013

Bergeron, F., Croteau, A., Uwizeyemungu, S., & Raymond, L. (2017). A Framework for Research on Information Technology Governance in SMEs. In S. De Haes & W. Van Grembergen (Eds.), *Strategic IT Governance and Alignment in Business Settings* (pp. 53–81). Hershey, PA: IGI Global. doi:10.4018/978-1-5225-0861-8.ch003

Bhardwaj, M., Shukla, N., & Sharma, A. (2021). Improvement and Reduction of Clustering Overhead in Mobile Ad Hoc Network With Optimum Stable Bunching Algorithm. In S. Kumar, M. Trivedi, P. Ranjan, & A. Punhani (Eds.), *Evolution of Software-Defined Networking Foundations for IoT and 5G Mobile Networks* (pp. 139–158). IGI Global. https://doi.org/10.4018/978-1-7998-4685-7.ch008

Bhatt, G. D., Wang, Z., & Rodger, J. A. (2017). Information Systems Capabilities and Their Effects on Competitive Advantages: A Study of Chinese Companies. *Information Resources Management Journal, 30*(3), 41–57. doi:10.4018/IRMJ.2017070103

Bhattacharya, A. (2021). Blockchain, Cybersecurity, and Industry 4.0. In A. Tyagi, G. Rekha, & N. Sreenath (Eds.), *Opportunities and Challenges for Blockchain Technology in Autonomous Vehicles* (pp. 210–244). IGI Global. https://doi.org/10.4018/978-1-7998-3295-9.ch013

Bhyan, P., Shrivastava, B., & Kumar, N. (2022). Requisite Sustainable Development Contemplating Buildings: Economic and Environmental Sustainability. In A. Hussain, K. Tiwari, & A. Gupta (Eds.), *Addressing Environmental Challenges Through Spatial Planning* (pp. 269–288). IGI Global. https://doi.org/10.4018/978-1-7998-8331-9.ch014

Boido, C., Davico, P., & Spallone, R. (2021). Digital Tools Aimed to Represent Urban Survey. In M. Khosrow-Pour D.B.A. (Ed.), *Encyclopedia of Information Science and Technology, Fifth Edition* (pp. 1181-1195). IGI Global. https://doi.org/10.4018/978-1-7998-3479-3.ch082

Borkar, P. S., Chanana, P. U., Atwal, S. K., Londe, T. G., & Dalal, Y. D. (2021). The Replacement of HMI (Human-Machine Interface) in Industry Using Single Interface Through IoT. In R. Raut & A. Mihovska (Eds.), *Examining the Impact of Deep Learning and IoT on Multi-Industry Applications* (pp. 195–208). IGI Global. https://doi.org/10.4018/978-1-7998-7511-6.ch011

Brahmane, A. V., & Krishna, C. B. (2021). Rider Chaotic Biography Optimization-driven Deep Stacked Auto-encoder for Big Data Classification Using Spark Architecture: Rider Chaotic Biography Optimization. *International Journal of Web Services Research, 18*(3), 42–62. https://doi.org/10.4018/ijwsr.2021070103

Burcoff, A., & Shamir, L. (2017). Computer Analysis of Pablo Picasso's Artistic Style. *International Journal of Art, Culture and Design Technologies, 6*(1), 1–18. doi:10.4018/IJACDT.2017010101

Byker, E. J. (2017). I Play I Learn: Introducing Technological Play Theory. In C. Martin & D. Polly (Eds.), *Handbook of Research on Teacher Education and Professional Development* (pp. 297–306). Hershey, PA: IGI Global. doi:10.4018/978-1-5225-1067-3.ch016

Calongne, C. M., Stricker, A. G., Truman, B., & Arenas, F. J. (2017). Cognitive Apprenticeship and Computer Science Education in Cyberspace: Reimagining the Past. In A. Stricker, C. Calongne, B. Truman, & F. Arenas (Eds.), *Integrating an Awareness of Selfhood and Society into Virtual Learning* (pp. 180–197). Hershey, PA: IGI Global. doi:10.4018/978-1-5225-2182-2.ch013

Carneiro, A. D. (2017). Defending Information Networks in Cyberspace: Some Notes on Security Needs. In M. Dawson, D. Kisku, P. Gupta, J. Sing, & W. Li (Eds.), Developing Next-Generation Countermeasures for Homeland Security Threat Prevention (pp. 354-375). Hershey, PA: IGI Global. https://doi.org/ doi:10.4018/978-1-5225-0703-1.ch016

Carvalho, W. F., & Zarate, L. (2021). Causal Feature Selection. In A. Azevedo & M. Santos (Eds.), *Integration Challenges for Analytics, Business Intelligence, and Data Mining* (pp. 145-160). IGI Global. https://doi.org/10.4018/978-1-7998-5781-5.ch007

Chase, J. P., & Yan, Z. (2017). Affect in Statistics Cognition. In *Assessing and Measuring Statistics Cognition in Higher Education Online Environments: Emerging Research and Opportunities* (pp. 144–187). Hershey, PA: IGI Global. doi:10.4018/978-1-5225-2420-5.ch005

Chatterjee, A., Roy, S., & Shrivastava, R. (2021). A Machine Learning Approach to Prevent Cancer. In G. Rani & P. Tiwari (Eds.), *Handbook of Research on Disease Prediction Through Data Analytics and Machine Learning* (pp. 112–141). IGI Global. https://doi.org/10.4018/978-1-7998-2742-9.ch007

Cifci, M. A. (2021). Optimizing WSNs for CPS Using Machine Learning Techniques. In A. Luhach & A. Elçi (Eds.), *Artificial Intelligence Paradigms for Smart Cyber-Physical Systems* (pp. 204–228). IGI Global. https://doi.org/10.4018/978-1-7998-5101-1.ch010

Cimermanova, I. (2017). Computer-Assisted Learning in Slovakia. In D. Tafazoli & M. Romero (Eds.), *Multiculturalism and Technology-Enhanced Language Learning* (pp. 252–270). Hershey, PA: IGI Global. doi:10.4018/978-1-5225-1882-2.ch015

Cipolla-Ficarra, F. V., & Cipolla-Ficarra, M. (2018). Computer Animation for Ingenious Revival. In F. Cipolla-Ficarra, M. Ficarra, M. Cipolla-Ficarra, A. Quiroga, J. Alma, & J. Carré (Eds.), *Technology-Enhanced Human Interaction in Modern Society* (pp. 159–181). Hershey, PA: IGI Global. doi:10.4018/978-1-5225-3437-2.ch008

Cockrell, S., Damron, T. S., Melton, A. M., & Smith, A. D. (2018). Offshoring IT. In M. Khosrow-Pour, D.B.A. (Ed.), Encyclopedia of Information Science and Technology, Fourth Edition (pp. 5476-5489). Hershey, PA: IGI Global. https://doi.org/ doi:10.4018/978-1-5225-2255-3.ch476

Coffey, J. W. (2018). Logic and Proof in Computer Science: Categories and Limits of Proof Techniques. In J. Horne (Ed.), *Philosophical Perceptions on Logic and Order* (pp. 218–240). Hershey, PA: IGI Global. doi:10.4018/978-1-5225-2443-4.ch007

Dale, M. (2017). Re-Thinking the Challenges of Enterprise Architecture Implementation. In M. Tavana (Ed.), *Enterprise Information Systems and the Digitalization of Business Functions* (pp. 205–221). Hershey, PA: IGI Global. doi:10.4018/978-1-5225-2382-6.ch009

Das, A., & Mohanty, M. N. (2021). An Useful Review on Optical Character Recognition for Smart Era Generation. In A. Tyagi (Ed.), *Multimedia and Sensory Input for Augmented, Mixed, and Virtual Reality* (pp. 1–41). IGI Global. https://doi.org/10.4018/978-1-7998-4703-8.ch001

Dash, A. K., & Mohapatra, P. (2021). A Survey on Prematurity Detection of Diabetic Retinopathy Based on Fundus Images Using Deep Learning Techniques. In S. Saxena & S. Paul (Eds.), *Deep Learning Applications in Medical Imaging* (pp. 140–155). IGI Global. https://doi.org/10.4018/978-1-7998-5071-7.ch006

De Maere, K., De Haes, S., & von Kutzschenbach, M. (2017). CIO Perspectives on Organizational Learning within the Context of IT Governance. *International Journal of IT/Business Alignment and Governance, 8*(1), 32-47. https://doi.org/doi:10.4018/IJITBAG.2017010103

Demir, K., Çaka, C., Yaman, N. D., İslamoğlu, H., & Kuzu, A. (2018). Examining the Current Definitions of Computational Thinking. In H. Ozcinar, G. Wong, & H. Ozturk (Eds.), *Teaching Computational Thinking in Primary Education* (pp. 36–64). Hershey, PA: IGI Global. doi:10.4018/978-1-5225-3200-2.ch003

Deng, X., Hung, Y., & Lin, C. D. (2017). Design and Analysis of Computer Experiments. In S. Saha, A. Mandal, A. Narasimhamurthy, S. V, & S. Sangam (Eds.), Handbook of Research on Applied Cybernetics and Systems Science (pp. 264-279). Hershey, PA: IGI Global. doi:10.4018/978-1-5225-2498-4.ch013

Denner, J., Martinez, J., & Thiry, H. (2017). Strategies for Engaging Hispanic/Latino Youth in the US in Computer Science. In Y. Rankin & J. Thomas (Eds.), *Moving Students of Color from Consumers to Producers of Technology* (pp. 24–48). Hershey, PA: IGI Global. doi:10.4018/978-1-5225-2005-4.ch002

Devi, A. (2017). Cyber Crime and Cyber Security: A Quick Glance. In R. Kumar, P. Pattnaik, & P. Pandey (Eds.), *Detecting and Mitigating Robotic Cyber Security Risks* (pp. 160–171). Hershey, PA: IGI Global. doi:10.4018/978-1-5225-2154-9.ch011

Dhaya, R., & Kanthavel, R. (2022). Futuristic Research Perspectives of IoT Platforms. In D. Jeya Mala (Ed.), *Integrating AI in IoT Analytics on the Cloud for Healthcare Applications* (pp. 258–275). IGI Global. doi:10.4018/978-1-7998-9132-1.ch015

Doyle, D. J., & Fahy, P. J. (2018). Interactivity in Distance Education and Computer-Aided Learning, With Medical Education Examples. In M. Khosrow-Pour, D.B.A. (Ed.), Encyclopedia of Information Science and Technology, Fourth Edition (pp. 5829-5840). Hershey, PA: IGI Global. https://doi.org/ doi:10.4018/978-1-5225-2255-3.ch507

Eklund, P. (2021). Reinforcement Learning in Social Media Marketing. In B. Christiansen & T. Škrinjarić (Eds.), *Handbook of Research on Applied AI for International Business and Marketing Applications* (pp. 30–48). IGI Global. https://doi.org/10.4018/978-1-7998-5077-9.ch003

El Ghandour, N., Benaissa, M., & Lebbah, Y. (2021). An Integer Linear Programming-Based Method for the Extraction of Ontology Alignment. *International Journal of Information Technology and Web Engineering*, *16*(2), 25–44. https://doi.org/10.4018/IJITWE.2021040102

Elias, N. I., & Walker, T. W. (2017). Factors that Contribute to Continued Use of E-Training among Healthcare Professionals. In F. Topor (Ed.), *Handbook of Research on Individualism and Identity in the Globalized Digital Age* (pp. 403–429). Hershey, PA: IGI Global. doi:10.4018/978-1-5225-0522-8.ch018

Fisher, R. L. (2018). Computer-Assisted Indian Matrimonial Services. In M. Khosrow-Pour, D.B.A. (Ed.), Encyclopedia of Information Science and Technology, Fourth Edition (pp. 4136-4145). Hershey, PA: IGI Global. doi:10.4018/978-1-5225-2255-3.ch358

Galiautdinov, R. (2021). Nonlinear Filtering in Artificial Neural Network Applications in Business and Engineering. In Q. Do (Ed.), *Artificial Neural Network Applications in Business and Engineering* (pp. 1–23). IGI Global. https://doi.org/10.4018/978-1-7998-3238-6.ch001

Gardner-McCune, C., & Jimenez, Y. (2017). Historical App Developers: Integrating CS into K-12 through Cross-Disciplinary Projects. In Y. Rankin & J. Thomas (Eds.), *Moving Students of Color from Consumers to Producers of Technology* (pp. 85–112). Hershey, PA: IGI Global. doi:10.4018/978-1-5225-2005-4.ch005

Garg, P. K. (2021). The Internet of Things-Based Technologies. In S. Kumar, M. Trivedi, P. Ranjan, & A. Punhani (Eds.), *Evolution of Software-Defined Networking Foundations for IoT and 5G Mobile Networks* (pp. 37–65). IGI Global. https://doi.org/10.4018/978-1-7998-4685-7.ch003

Garg, T., & Bharti, M. (2021). Congestion Control Protocols for UWSNs. In N. Goyal, L. Sapra, & J. Sandhu (Eds.), *Energy-Efficient Underwater Wireless Communications and Networking* (pp. 85–100). IGI Global. https://doi.org/10.4018/978-1-7998-3640-7.ch006

Gauttier, S. (2021). A Primer on Q-Method and the Study of Technology. In M. Khosrow-Pour D.B.A. (Eds.), *Encyclopedia of Information Science and Technology, Fifth Edition* (pp. 1746-1756). IGI Global. https://doi.org/10.4018/978-1-7998-3479-3.ch120

Ghafele, R., & Gibert, B. (2018). Open Growth: The Economic Impact of Open Source Software in the USA. In M. Khosrow-Pour (Ed.), *Optimizing Contemporary Application and Processes in Open Source Software* (pp. 164–197). Hershey, PA: IGI Global. doi:10.4018/978-1-5225-5314-4.ch007

Ghobakhloo, M., & Azar, A. (2018). Information Technology Resources, the Organizational Capability of Lean-Agile Manufacturing, and Business Performance. *Information Resources Management Journal, 31*(2), 47–74. doi:10.4018/IRMJ.2018040103

Gikandi, J. W. (2017). Computer-Supported Collaborative Learning and Assessment: A Strategy for Developing Online Learning Communities in Continuing Education. In J. Keengwe & G. Onchwari (Eds.), *Handbook of Research on Learner-Centered Pedagogy in Teacher Education and Professional Development* (pp. 309–333). Hershey, PA: IGI Global. doi:10.4018/978-1-5225-0892-2.ch017

Gokhale, A. A., & Machina, K. F. (2017). Development of a Scale to Measure Attitudes toward Information Technology. In L. Tomei (Ed.), *Exploring the New Era of Technology-Infused Education* (pp. 49–64). Hershey, PA: IGI Global. doi:10.4018/978-1-5225-1709-2.ch004

Goswami, J. K., Jalal, S., Negi, C. S., & Jalal, A. S. (2022). A Texture Features-Based Robust Facial Expression Recognition. *International Journal of Computer Vision and Image Processing, 12*(1), 1–15. https://doi.org/10.4018/IJCVIP.2022010103

Hafeez-Baig, A., Gururajan, R., & Wickramasinghe, N. (2017). Readiness as a Novel Construct of Readiness Acceptance Model (RAM) for the Wireless Handheld Technology. In N. Wickramasinghe (Ed.), *Handbook of Research on Healthcare Administration and Management* (pp. 578–595). Hershey, PA: IGI Global. doi:10.4018/978-1-5225-0920-2.ch035

Hanafizadeh, P., Ghandchi, S., & Asgarimehr, M. (2017). Impact of Information Technology on Lifestyle: A Literature Review and Classification. *International Journal of Virtual Communities and Social Networking*, *9*(2), 1–23. doi:10.4018/IJVCSN.2017040101

Haseski, H. İ., Ilic, U., & Tuğtekin, U. (2018). Computational Thinking in Educational Digital Games: An Assessment Tool Proposal. In H. Ozcinar, G. Wong, & H. Ozturk (Eds.), *Teaching Computational Thinking in Primary Education* (pp. 256–287). Hershey, PA: IGI Global. doi:10.4018/978-1-5225-3200-2.ch013

Hee, W. J., Jalleh, G., Lai, H., & Lin, C. (2017). E-Commerce and IT Projects: Evaluation and Management Issues in Australian and Taiwanese Hospitals. *International Journal of Public Health Management and Ethics*, *2*(1), 69–90. doi:10.4018/IJPHME.2017010104

Hernandez, A. A. (2017). Green Information Technology Usage: Awareness and Practices of Philippine IT Professionals. *International Journal of Enterprise Information Systems*, *13*(4), 90–103. doi:10.4018/IJEIS.2017100106

Hernandez, M. A., Marin, E. C., Garcia-Rodriguez, J., Azorin-Lopez, J., & Cazorla, M. (2017). Automatic Learning Improves Human-Robot Interaction in Productive Environments: A Review. *International Journal of Computer Vision and Image Processing*, *7*(3), 65–75. doi:10.4018/IJCVIP.2017070106

Hirota, A. (2021). Design of Narrative Creation in Innovation: "Signature Story" and Two Types of Pivots. In T. Ogata & J. Ono (Eds.), *Bridging the Gap Between AI, Cognitive Science, and Narratology With Narrative Generation* (pp. 363–376). IGI Global. https://doi.org/10.4018/978-1-7998-4864-6.ch012

Hond, D., Asgari, H., Jeffery, D., & Newman, M. (2021). An Integrated Process for Verifying Deep Learning Classifiers Using Dataset Dissimilarity Measures. *International Journal of Artificial Intelligence and Machine Learning*, *11*(2), 1–21. https://doi.org/10.4018/IJAIML.289536

Horne-Popp, L. M., Tessone, E. B., & Welker, J. (2018). If You Build It, They Will Come: Creating a Library Statistics Dashboard for Decision-Making. In L. Costello & M. Powers (Eds.), *Developing In-House Digital Tools in Library Spaces* (pp. 177–203). Hershey, PA: IGI Global. doi:10.4018/978-1-5225-2676-6.ch009

Hu, H., Hu, P. J., & Al-Gahtani, S. S. (2017). User Acceptance of Computer Technology at Work in Arabian Culture: A Model Comparison Approach. In M. Khosrow-Pour (Ed.), *Handbook of Research on Technology Adoption, Social Policy, and Global Integration* (pp. 205–228). Hershey, PA: IGI Global. doi:10.4018/978-1-5225-2668-1.ch011

Huang, C., Sun, Y., & Fuh, C. (2022). Vehicle License Plate Recognition With Deep Learning. In C. Chen, W. Yang, & L. Chen (Eds.), *Technologies to Advance Automation in Forensic Science and Criminal Investigation* (pp. 161-219). IGI Global. https://doi.org/10.4018/978-1-7998-8386-9.ch009

Ifinedo, P. (2017). Using an Extended Theory of Planned Behavior to Study Nurses' Adoption of Healthcare Information Systems in Nova Scotia. *International Journal of Technology Diffusion*, *8*(1), 1–17. doi:10.4018/IJTD.2017010101

Ilie, V., & Sneha, S. (2018). A Three Country Study for Understanding Physicians' Engagement With Electronic Information Resources Pre and Post System Implementation. *Journal of Global Information Management*, *26*(2), 48–73. doi:10.4018/JGIM.2018040103

Ilo, P. I., Nkiko, C., Ugwu, C. I., Ekere, J. N., Izuagbe, R., & Fagbohun, M. O. (2021). Prospects and Challenges of Web 3.0 Technologies Application in the Provision of Library Services. In M. Khosrow-Pour D.B.A. (Ed.), *Encyclopedia of Information Science and Technology, Fifth Edition* (pp. 1767-1781). IGI Global. https://doi.org/10.4018/978-1-7998-3479-3.ch122

Inoue-Smith, Y. (2017). Perceived Ease in Using Technology Predicts Teacher Candidates' Preferences for Online Resources. *International Journal of Online Pedagogy and Course Design*, *7*(3), 17–28. doi:10.4018/IJOPCD.2017070102

Islam, A. Y. (2017). Technology Satisfaction in an Academic Context: Moderating Effect of Gender. In A. Mesquita (Ed.), *Research Paradigms and Contemporary Perspectives on Human-Technology Interaction* (pp. 187–211). Hershey, PA: IGI Global. doi:10.4018/978-1-5225-1868-6.ch009

Jagdale, S. C., Hable, A. A., & Chabukswar, A. R. (2021). Protocol Development in Clinical Trials for Healthcare Management. In M. Khosrow-Pour D.B.A. (Ed.), *Encyclopedia of Information Science and Technology, Fifth Edition* (pp. 1797-1814). IGI Global. https://doi.org/10.4018/978-1-7998-3479-3.ch124

Jamil, G. L., & Jamil, C. C. (2017). Information and Knowledge Management Perspective Contributions for Fashion Studies: Observing Logistics and Supply Chain Management Processes. In G. Jamil, A. Soares, & C. Pessoa (Eds.), *Handbook of Research on Information Management for Effective Logistics and Supply Chains* (pp. 199–221). Hershey, PA: IGI Global. doi:10.4018/978-1-5225-0973-8.ch011

Jamil, M. I., & Almunawar, M. N. (2021). Importance of Digital Literacy and Hindrance Brought About by Digital Divide. In M. Khosrow-Pour D.B.A. (Ed.), *Encyclopedia of Information Science and Technology, Fifth Edition* (pp. 1683-1698). IGI Global. https://doi.org/10.4018/978-1-7998-3479-3.ch116

Janakova, M. (2018). Big Data and Simulations for the Solution of Controversies in Small Businesses. In M. Khosrow-Pour, D.B.A. (Ed.), Encyclopedia of Information Science and Technology, Fourth Edition (pp. 6907-6915). Hershey, PA: IGI Global. doi:10.4018/978-1-5225-2255-3.ch598

Jhawar, A., & Garg, S. K. (2018). Logistics Improvement by Investment in Information Technology Using System Dynamics. In A. Azar & S. Vaidyanathan (Eds.), *Advances in System Dynamics and Control* (pp. 528–567). Hershey, PA: IGI Global. doi:10.4018/978-1-5225-4077-9.ch017

Kalelioğlu, F., Gülbahar, Y., & Doğan, D. (2018). Teaching How to Think Like a Programmer: Emerging Insights. In H. Ozcinar, G. Wong, & H. Ozturk (Eds.), *Teaching Computational Thinking in Primary Education* (pp. 18–35). Hershey, PA: IGI Global. doi:10.4018/978-1-5225-3200-2.ch002

Kamberi, S. (2017). A Girls-Only Online Virtual World Environment and its Implications for Game-Based Learning. In A. Stricker, C. Calongne, B. Truman, & F. Arenas (Eds.), *Integrating an Awareness of Selfhood and Society into Virtual Learning* (pp. 74–95). Hershey, PA: IGI Global. doi:10.4018/978-1-5225-2182-2.ch006

Kamel, S., & Rizk, N. (2017). ICT Strategy Development: From Design to Implementation – Case of Egypt. In C. Howard & K. Hargiss (Eds.), *Strategic Information Systems and Technologies in Modern Organizations* (pp. 239–257). Hershey, PA: IGI Global. doi:10.4018/978-1-5225-1680-4.ch010

Kamel, S. H. (2018). The Potential Role of the Software Industry in Supporting Economic Development. In M. Khosrow-Pour, D.B.A. (Ed.), Encyclopedia of Information Science and Technology, Fourth Edition (pp. 7259-7269). Hershey, PA: IGI Global. doi:10.4018/978-1-5225-2255-3.ch631

Kang, H., Kang, Y., & Kim, J. (2022). Improved Fall Detection Model on GRU Using PoseNet. *International Journal of Software Innovation, 10*(2), 1–11. https://doi.org/10.4018/IJSI.289600

Kankam, P. K. (2021). Employing Case Study and Survey Designs in Information Research. *Journal of Information Technology Research, 14*(1), 167–177. https://doi.org/10.4018/JITR.2021010110

Karas, V., & Schuller, B. W. (2021). Deep Learning for Sentiment Analysis: An Overview and Perspectives. In F. Pinarbasi & M. Taskiran (Eds.), *Natural Language Processing for Global and Local Business* (pp. 97–132). IGI Global. https://doi.org/10.4018/978-1-7998-4240-8.ch005

Kaufman, L. M. (2022). Reimagining the Magic of the Workshop Model. In T. Driscoll III, (Ed.), *Designing Effective Distance and Blended Learning Environments in K-12* (pp. 89–109). IGI Global. https://doi.org/10.4018/978-1-7998-6829-3.ch007

Kawata, S. (2018). Computer-Assisted Parallel Program Generation. In M. Khosrow-Pour, D.B.A. (Ed.), Encyclopedia of Information Science and Technology, Fourth Edition (pp. 4583-4593). Hershey, PA: IGI Global. doi:10.4018/978-1-5225-2255-3.ch398

Kharb, L., & Singh, P. (2021). Role of Machine Learning in Modern Education and Teaching. In S. Verma & P. Tomar (Ed.), *Impact of AI Technologies on Teaching, Learning, and Research in Higher Education* (pp. 99-123). IGI Global. https://doi.org/10.4018/978-1-7998-4763-2.ch006

Khari, M., Shrivastava, G., Gupta, S., & Gupta, R. (2017). Role of Cyber Security in Today's Scenario. In R. Kumar, P. Pattnaik, & P. Pandey (Eds.), *Detecting and Mitigating Robotic Cyber Security Risks* (pp. 177–191). Hershey, PA: IGI Global. doi:10.4018/978-1-5225-2154-9.ch013

Khekare, G., & Sheikh, S. (2021). Autonomous Navigation Using Deep Reinforcement Learning in ROS. *International Journal of Artificial Intelligence and Machine Learning*, *11*(2), 63–70. https://doi.org/10.4018/IJAIML.20210701.oa4

Khouja, M., Rodriguez, I. B., Ben Halima, Y., & Moalla, S. (2018). IT Governance in Higher Education Institutions: A Systematic Literature Review. *International Journal of Human Capital and Information Technology Professionals*, *9*(2), 52–67. doi:10.4018/IJHCITP.2018040104

Kiourt, C., Pavlidis, G., Koutsoudis, A., & Kalles, D. (2017). Realistic Simulation of Cultural Heritage. *International Journal of Computational Methods in Heritage Science*, *1*(1), 10–40. doi:10.4018/IJCMHS.2017010102

Köse, U. (2017). An Augmented-Reality-Based Intelligent Mobile Application for Open Computer Education. In G. Kurubacak & H. Altinpulluk (Eds.), *Mobile Technologies and Augmented Reality in Open Education* (pp. 154–174). Hershey, PA: IGI Global. doi:10.4018/978-1-5225-2110-5.ch008

Lahmiri, S. (2018). Information Technology Outsourcing Risk Factors and Provider Selection. In M. Gupta, R. Sharman, J. Walp, & P. Mulgund (Eds.), *Information Technology Risk Management and Compliance in Modern Organizations* (pp. 214–228). Hershey, PA: IGI Global. doi:10.4018/978-1-5225-2604-9.ch008

Lakkad, A. K., Bhadaniya, R. D., Shah, V. N., & Lavanya, K. (2021). Complex Events Processing on Live News Events Using Apache Kafka and Clustering Techniques. *International Journal of Intelligent Information Technologies, 17*(1), 39–52. https://doi.org/10.4018/IJIIT.2021010103

Landriscina, F. (2017). Computer-Supported Imagination: The Interplay Between Computer and Mental Simulation in Understanding Scientific Concepts. In I. Levin & D. Tsybulsky (Eds.), *Digital Tools and Solutions for Inquiry-Based STEM Learning* (pp. 33–60). Hershey, PA: IGI Global. doi:10.4018/978-1-5225-2525-7.ch002

Lara López, G. (2021). Virtual Reality in Object Location. In A. Negrón & M. Muñoz (Eds.), *Latin American Women and Research Contributions to the IT Field* (pp. 307–324). IGI Global. https://doi.org/10.4018/978-1-7998-7552-9.ch014

Lee, W. W. (2018). Ethical Computing Continues From Problem to Solution. In M. Khosrow-Pour, D.B.A. (Ed.), Encyclopedia of Information Science and Technology, Fourth Edition (pp. 4884-4897). Hershey, PA: IGI Global. doi:10.4018/978-1-5225-2255-3.ch423

Lin, S., Chen, S., & Chuang, S. (2017). Perceived Innovation and Quick Response Codes in an Online-to-Offline E-Commerce Service Model. *International Journal of E-Adoption, 9*(2), 1–16. doi:10.4018/IJEA.2017070101

Liu, M., Wang, Y., Xu, W., & Liu, L. (2017). Automated Scoring of Chinese Engineering Students' English Essays. *International Journal of Distance Education Technologies, 15*(1), 52–68. doi:10.4018/IJDET.2017010104

Ma, X., Li, X., Zhong, B., Huang, Y., Gu, Y., Wu, M., Liu, Y., & Zhang, M. (2021). A Detector and Evaluation Framework of Abnormal Bidding Behavior Based on Supplier Portrait. *International Journal of Information Technology and Web Engineering, 16*(2), 58–74. https://doi.org/10.4018/IJITWE.2021040104

Mabe, L. K., & Oladele, O. I. (2017). Application of Information Communication Technologies for Agricultural Development through Extension Services: A Review. In T. Tossy (Ed.), *Information Technology Integration for Socio-Economic Development* (pp. 52–101). Hershey, PA: IGI Global. doi:10.4018/978-1-5225-0539-6.ch003

Mahboub, S. A., Sayed Ali Ahmed, E., & Saeed, R. A. (2021). Smart IDS and IPS for Cyber-Physical Systems. In A. Luhach & A. Elçi (Eds.), *Artificial Intelligence Paradigms for Smart Cyber-Physical Systems* (pp. 109–136). IGI Global. https://doi.org/10.4018/978-1-7998-5101-1.ch006

Manogaran, G., Thota, C., & Lopez, D. (2018). Human-Computer Interaction With Big Data Analytics. In D. Lopez & M. Durai (Eds.), *HCI Challenges and Privacy Preservation in Big Data Security* (pp. 1–22). Hershey, PA: IGI Global. doi:10.4018/978-1-5225-2863-0.ch001

Margolis, J., Goode, J., & Flapan, J. (2017). A Critical Crossroads for Computer Science for All: "Identifying Talent" or "Building Talent," and What Difference Does It Make? In Y. Rankin & J. Thomas (Eds.), *Moving Students of Color from Consumers to Producers of Technology* (pp. 1–23). Hershey, PA: IGI Global. doi:10.4018/978-1-5225-2005-4.ch001

Mazzù, M. F., Benetton, A., Baccelloni, A., & Lavini, L. (2022). A Milk Blockchain-Enabled Supply Chain: Evidence From Leading Italian Farms. In P. De Giovanni (Ed.), *Blockchain Technology Applications in Businesses and Organizations* (pp. 73–98). IGI Global. https://doi.org/10.4018/978-1-7998-8014-1.ch004

Mbale, J. (2018). Computer Centres Resource Cloud Elasticity-Scalability (CRECES): Copperbelt University Case Study. In S. Aljawarneh & M. Malhotra (Eds.), *Critical Research on Scalability and Security Issues in Virtual Cloud Environments* (pp. 48–70). Hershey, PA: IGI Global. doi:10.4018/978-1-5225-3029-9.ch003

McKee, J. (2018). The Right Information: The Key to Effective Business Planning. In *Business Architectures for Risk Assessment and Strategic Planning: Emerging Research and Opportunities* (pp. 38–52). Hershey, PA: IGI Global. doi:10.4018/978-1-5225-3392-4.ch003

Meddah, I. H., Remil, N. E., & Meddah, H. N. (2021). Novel Approach for Mining Patterns. *International Journal of Applied Evolutionary Computation, 12*(1), 27–42. https://doi.org/10.4018/IJAEC.2021010103

Mensah, I. K., & Mi, J. (2018). Determinants of Intention to Use Local E-Government Services in Ghana: The Perspective of Local Government Workers. *International Journal of Technology Diffusion, 9*(2), 41–60. doi:10.4018/IJTD.2018040103

Mohamed, J. H. (2018). Scientograph-Based Visualization of Computer Forensics Research Literature. In J. Jeyasekar & P. Saravanan (Eds.), *Innovations in Measuring and Evaluating Scientific Information* (pp. 148–162). Hershey, PA: IGI Global. doi:10.4018/978-1-5225-3457-0.ch010

Montañés-Del Río, M. Á., Cornejo, V. R., Rodríguez, M. R., & Ortiz, J. S. (2021). Gamification of University Subjects: A Case Study for Operations Management. *Journal of Information Technology Research, 14*(2), 1–29. https://doi.org/10.4018/JITR.2021040101

Moore, R. L., & Johnson, N. (2017). Earning a Seat at the Table: How IT Departments Can Partner in Organizational Change and Innovation. *International Journal of Knowledge-Based Organizations*, *7*(2), 1–12. doi:10.4018/IJKBO.2017040101

Mukul, M. K., & Bhattaharyya, S. (2017). Brain-Machine Interface: Human-Computer Interaction. In E. Noughabi, B. Raahemi, A. Albadvi, & B. Far (Eds.), *Handbook of Research on Data Science for Effective Healthcare Practice and Administration* (pp. 417–443). Hershey, PA: IGI Global. doi:10.4018/978-1-5225-2515-8.ch018

Na, L. (2017). Library and Information Science Education and Graduate Programs in Academic Libraries. In L. Ruan, Q. Zhu, & Y. Ye (Eds.), *Academic Library Development and Administration in China* (pp. 218–229). Hershey, PA: IGI Global. doi:10.4018/978-1-5225-0550-1.ch013

Nagpal, G., Bishnoi, G. K., Dhami, H. S., & Vijayvargia, A. (2021). Use of Data Analytics to Increase the Efficiency of Last Mile Logistics for Ecommerce Deliveries. In B. Patil & M. Vohra (Eds.), *Handbook of Research on Engineering, Business, and Healthcare Applications of Data Science and Analytics* (pp. 167–180). IGI Global. https://doi.org/10.4018/978-1-7998-3053-5.ch009

Nair, S. M., Ramesh, V., & Tyagi, A. K. (2021). Issues and Challenges (Privacy, Security, and Trust) in Blockchain-Based Applications. In A. Tyagi, G. Rekha, & N. Sreenath (Eds.), *Opportunities and Challenges for Blockchain Technology in Autonomous Vehicles* (pp. 196–209). IGI Global. https://doi.org/10.4018/978-1-7998-3295-9.ch012

Naomi, J. F. M., K., & V., S. (2021). Machine and Deep Learning Techniques in IoT and Cloud. In S. Velayutham (Ed.), *Challenges and Opportunities for the Convergence of IoT, Big Data, and Cloud Computing* (pp. 225-247). IGI Global. https://doi.org/10.4018/978-1-7998-3111-2.ch013

Nath, R., & Murthy, V. N. (2018). What Accounts for the Differences in Internet Diffusion Rates Around the World? In M. Khosrow-Pour, D.B.A. (Ed.), Encyclopedia of Information Science and Technology, Fourth Edition (pp. 8095-8104). Hershey, PA: IGI Global. https://doi.org/ doi:10.4018/978-1-5225-2255-3.ch705

Nedelko, Z., & Potocan, V. (2018). The Role of Emerging Information Technologies for Supporting Supply Chain Management. In M. Khosrow-Pour, D.B.A. (Ed.), Encyclopedia of Information Science and Technology, Fourth Edition (pp. 5559-5569). Hershey, PA: IGI Global. doi:10.4018/978-1-5225-2255-3.ch483

Negrini, L., Giang, C., & Bonnet, E. (2022). Designing Tools and Activities for Educational Robotics in Online Learning. In N. Eteokleous & E. Nisiforou (Eds.), *Designing, Constructing, and Programming Robots for Learning* (pp. 202–222). IGI Global. https://doi.org/10.4018/978-1-7998-7443-0.ch010

Ngafeeson, M. N. (2018). User Resistance to Health Information Technology. In M. Khosrow-Pour, D.B.A. (Ed.), Encyclopedia of Information Science and Technology, Fourth Edition (pp. 3816-3825). Hershey, PA: IGI Global. doi:10.4018/978-1-5225-2255-3.ch331

Nguyen, T. T., Giang, N. L., Tran, D. T., Nguyen, T. T., Nguyen, H. Q., Pham, A. V., & Vu, T. D. (2021). A Novel Filter-Wrapper Algorithm on Intuitionistic Fuzzy Set for Attribute Reduction From Decision Tables. *International Journal of Data Warehousing and Mining, 17*(4), 67–100. https://doi.org/10.4018/IJDWM.2021100104

Nigam, A., & Dewani, P. P. (2022). Consumer Engagement Through Conditional Promotions: An Exploratory Study. *Journal of Global Information Management, 30*(5), 1–19. https://doi.org/10.4018/JGIM.290364

Odagiri, K. (2017). Introduction of Individual Technology to Constitute the Current Internet. In *Strategic Policy-Based Network Management in Contemporary Organizations* (pp. 20–96). Hershey, PA: IGI Global. doi:10.4018/978-1-68318-003-6.ch003

Odia, J. O., & Akpata, O. T. (2021). Role of Data Science and Data Analytics in Forensic Accounting and Fraud Detection. In B. Patil & M. Vohra (Eds.), *Handbook of Research on Engineering, Business, and Healthcare Applications of Data Science and Analytics* (pp. 203–227). IGI Global. https://doi.org/10.4018/978-1-7998-3053-5.ch011

Okike, E. U. (2018). Computer Science and Prison Education. In I. Biao (Ed.), *Strategic Learning Ideologies in Prison Education Programs* (pp. 246–264). Hershey, PA: IGI Global. doi:10.4018/978-1-5225-2909-5.ch012

Olelewe, C. J., & Nwafor, I. P. (2017). Level of Computer Appreciation Skills Acquired for Sustainable Development by Secondary School Students in Nsukka LGA of Enugu State, Nigeria. In C. Ayo & V. Mbarika (Eds.), *Sustainable ICT Adoption and Integration for Socio-Economic Development* (pp. 214–233). Hershey, PA: IGI Global. doi:10.4018/978-1-5225-2565-3.ch010

Oliveira, M., Maçada, A. C., Curado, C., & Nodari, F. (2017). Infrastructure Profiles and Knowledge Sharing. *International Journal of Technology and Human Interaction, 13*(3), 1–12. doi:10.4018/IJTHI.2017070101

Otarkhani, A., Shokouhyar, S., & Pour, S. S. (2017). Analyzing the Impact of Governance of Enterprise IT on Hospital Performance: Tehran's (Iran) Hospitals – A Case Study. *International Journal of Healthcare Information Systems and Informatics*, *12*(3), 1–20. doi:10.4018/IJHISI.2017070101

Otunla, A. O., & Amuda, C. O. (2018). Nigerian Undergraduate Students' Computer Competencies and Use of Information Technology Tools and Resources for Study Skills and Habits' Enhancement. In M. Khosrow-Pour, D.B.A. (Ed.), Encyclopedia of Information Science and Technology, Fourth Edition (pp. 2303-2313). Hershey, PA: IGI Global. https://doi.org/ doi:10.4018/978-1-5225-2255-3.ch200

Özçınar, H. (2018). A Brief Discussion on Incentives and Barriers to Computational Thinking Education. In H. Ozcinar, G. Wong, & H. Ozturk (Eds.), *Teaching Computational Thinking in Primary Education* (pp. 1–17). Hershey, PA: IGI Global. doi:10.4018/978-1-5225-3200-2.ch001

Pandey, J. M., Garg, S., Mishra, P., & Mishra, B. P. (2017). Computer Based Psychological Interventions: Subject to the Efficacy of Psychological Services. *International Journal of Computers in Clinical Practice*, *2*(1), 25–33. doi:10.4018/IJCCP.2017010102

Pandkar, S. D., & Paatil, S. D. (2021). Big Data and Knowledge Resource Centre. In S. Dhamdhere (Ed.), *Big Data Applications for Improving Library Services* (pp. 90–106). IGI Global. https://doi.org/10.4018/978-1-7998-3049-8.ch007

Patro, C. (2017). Impulsion of Information Technology on Human Resource Practices. In P. Ordóñez de Pablos (Ed.), *Managerial Strategies and Solutions for Business Success in Asia* (pp. 231–254). Hershey, PA: IGI Global. doi:10.4018/978-1-5225-1886-0.ch013

Patro, C. S., & Raghunath, K. M. (2017). Information Technology Paraphernalia for Supply Chain Management Decisions. In M. Tavana (Ed.), *Enterprise Information Systems and the Digitalization of Business Functions* (pp. 294–320). Hershey, PA: IGI Global. doi:10.4018/978-1-5225-2382-6.ch014

Paul, P. K. (2018). The Context of IST for Solid Information Retrieval and Infrastructure Building: Study of Developing Country. *International Journal of Information Retrieval Research*, *8*(1), 86–100. doi:10.4018/IJIRR.2018010106

Paul, P. K., & Chatterjee, D. (2018). iSchools Promoting "Information Science and Technology" (IST) Domain Towards Community, Business, and Society With Contemporary Worldwide Trend and Emerging Potentialities in India. In M. Khosrow-Pour, D.B.A. (Ed.), Encyclopedia of Information Science and Technology, Fourth Edition (pp. 4723-4735). Hershey, PA: IGI Global. https://doi.org/ doi:10.4018/978-1-5225-2255-3.ch410

Pessoa, C. R., & Marques, M. E. (2017). Information Technology and Communication Management in Supply Chain Management. In G. Jamil, A. Soares, & C. Pessoa (Eds.), *Handbook of Research on Information Management for Effective Logistics and Supply Chains* (pp. 23–33). Hershey, PA: IGI Global. doi:10.4018/978-1-5225-0973-8.ch002

Pineda, R. G. (2018). Remediating Interaction: Towards a Philosophy of Human-Computer Relationship. In M. Khosrow-Pour (Ed.), *Enhancing Art, Culture, and Design With Technological Integration* (pp. 75–98). Hershey, PA: IGI Global. doi:10.4018/978-1-5225-5023-5.ch004

Prabha, V. D., & R., R. (2021). Clinical Decision Support Systems: Decision-Making System for Clinical Data. In G. Rani & P. Tiwari (Eds.), *Handbook of Research on Disease Prediction Through Data Analytics and Machine Learning* (pp. 268-280). IGI Global. https://doi.org/10.4018/978-1-7998-2742-9.ch014

Pushpa, R., & Siddappa, M. (2021). An Optimal Way of VM Placement Strategy in Cloud Computing Platform Using ABCS Algorithm. *International Journal of Ambient Computing and Intelligence*, *12*(3), 16–38. https://doi.org/10.4018/IJACI.2021070102

Qian, Y. (2017). Computer Simulation in Higher Education: Affordances, Opportunities, and Outcomes. In P. Vu, S. Fredrickson, & C. Moore (Eds.), *Handbook of Research on Innovative Pedagogies and Technologies for Online Learning in Higher Education* (pp. 236–262). Hershey, PA: IGI Global. doi:10.4018/978-1-5225-1851-8.ch011

Rahman, N. (2017). Lessons from a Successful Data Warehousing Project Management. *International Journal of Information Technology Project Management*, *8*(4), 30–45. doi:10.4018/IJITPM.2017100103

Rahman, N. (2018). Environmental Sustainability in the Computer Industry for Competitive Advantage. In M. Khosrow-Pour (Ed.), *Green Computing Strategies for Competitive Advantage and Business Sustainability* (pp. 110–130). Hershey, PA: IGI Global. doi:10.4018/978-1-5225-5017-4.ch006

Rajh, A., & Pavetic, T. (2017). Computer Generated Description as the Required Digital Competence in Archival Profession. *International Journal of Digital Literacy and Digital Competence*, *8*(1), 36–49. doi:10.4018/IJDLDC.2017010103

Raman, A., & Goyal, D. P. (2017). Extending IMPLEMENT Framework for Enterprise Information Systems Implementation to Information System Innovation. In M. Tavana (Ed.), *Enterprise Information Systems and the Digitalization of Business Functions* (pp. 137–177). Hershey, PA: IGI Global. doi:10.4018/978-1-5225-2382-6.ch007

Rao, A. P., & Reddy, K. S. (2021). Automated Soil Residue Levels Detecting Device With IoT Interface. In V. Sathiyamoorthi & A. Elci (Eds.), *Challenges and Applications of Data Analytics in Social Perspectives* (Vol. S, pp. 123–135). IGI Global. https://doi.org/10.4018/978-1-7998-2566-1.ch007

Rao, Y. S., Rauta, A. K., Saini, H., & Panda, T. C. (2017). Mathematical Model for Cyber Attack in Computer Network. *International Journal of Business Data Communications and Networking*, *13*(1), 58–65. doi:10.4018/IJBDCN.2017010105

Rapaport, W. J. (2018). Syntactic Semantics and the Proper Treatment of Computationalism. In M. Danesi (Ed.), *Empirical Research on Semiotics and Visual Rhetoric* (pp. 128–176). Hershey, PA: IGI Global. doi:10.4018/978-1-5225-5622-0.ch007

Raut, R., Priyadarshinee, P., & Jha, M. (2017). Understanding the Mediation Effect of Cloud Computing Adoption in Indian Organization: Integrating TAM-TOE- Risk Model. *International Journal of Service Science, Management, Engineering, and Technology*, *8*(3), 40–59. doi:10.4018/IJSSMET.2017070103

Rezaie, S., Mirabedini, S. J., & Abtahi, A. (2018). Designing a Model for Implementation of Business Intelligence in the Banking Industry. *International Journal of Enterprise Information Systems*, *14*(1), 77–103. doi:10.4018/IJEIS.2018010105

Rezende, D. A. (2018). Strategic Digital City Projects: Innovative Information and Public Services Offered by Chicago (USA) and Curitiba (Brazil). In M. Lytras, L. Daniela, & A. Visvizi (Eds.), *Enhancing Knowledge Discovery and Innovation in the Digital Era* (pp. 204–223). Hershey, PA: IGI Global. doi:10.4018/978-1-5225-4191-2.ch012

Rodriguez, A., Rico-Diaz, A. J., Rabuñal, J. R., & Gestal, M. (2017). Fish Tracking with Computer Vision Techniques: An Application to Vertical Slot Fishways. In M. S., & V. V. (Eds.), Multi-Core Computer Vision and Image Processing for Intelligent Applications (pp. 74-104). Hershey, PA: IGI Global. https://doi.org/doi:10.4018/978-1-5225-0889-2.ch003

Romero, J. A. (2018). Sustainable Advantages of Business Value of Information Technology. In M. Khosrow-Pour, D.B.A. (Ed.), Encyclopedia of Information Science and Technology, Fourth Edition (pp. 923-929). Hershey, PA: IGI Global. doi:10.4018/978-1-5225-2255-3.ch079

Romero, J. A. (2018). The Always-On Business Model and Competitive Advantage. In N. Bajgoric (Ed.), *Always-On Enterprise Information Systems for Modern Organizations* (pp. 23–40). Hershey, PA: IGI Global. doi:10.4018/978-1-5225-3704-5.ch002

Rosen, Y. (2018). Computer Agent Technologies in Collaborative Learning and Assessment. In M. Khosrow-Pour, D.B.A. (Ed.), Encyclopedia of Information Science and Technology, Fourth Edition (pp. 2402-2410). Hershey, PA: IGI Global. doi:10.4018/978-1-5225-2255-3.ch209

Roy, D. (2018). Success Factors of Adoption of Mobile Applications in Rural India: Effect of Service Characteristics on Conceptual Model. In M. Khosrow-Pour (Ed.), *Green Computing Strategies for Competitive Advantage and Business Sustainability* (pp. 211–238). Hershey, PA: IGI Global. doi:10.4018/978-1-5225-5017-4.ch010

Ruffin, T. R., & Hawkins, D. P. (2018). Trends in Health Care Information Technology and Informatics. In M. Khosrow-Pour, D.B.A. (Ed.), Encyclopedia of Information Science and Technology, Fourth Edition (pp. 3805-3815). Hershey, PA: IGI Global. doi:10.4018/978-1-5225-2255-3.ch330

Sadasivam, U. M., & Ganesan, N. (2021). Detecting Fake News Using Deep Learning and NLP. In S. Misra, C. Arumugam, S. Jaganathan, & S. S. (Eds.), *Confluence of AI, Machine, and Deep Learning in Cyber Forensics* (pp. 117-133). IGI Global. https://doi.org/10.4018/978-1-7998-4900-1.ch007

Safari, M. R., & Jiang, Q. (2018). The Theory and Practice of IT Governance Maturity and Strategies Alignment: Evidence From Banking Industry. *Journal of Global Information Management*, *26*(2), 127–146. doi:10.4018/JGIM.2018040106

Sahin, H. B., & Anagun, S. S. (2018). Educational Computer Games in Math Teaching: A Learning Culture. In E. Toprak & E. Kumtepe (Eds.), *Supporting Multiculturalism in Open and Distance Learning Spaces* (pp. 249–280). Hershey, PA: IGI Global. doi:10.4018/978-1-5225-3076-3.ch013

Sakalle, A., Tomar, P., Bhardwaj, H., & Sharma, U. (2021). Impact and Latest Trends of Intelligent Learning With Artificial Intelligence. In S. Verma & P. Tomar (Eds.), *Impact of AI Technologies on Teaching, Learning, and Research in Higher Education* (pp. 172-189). IGI Global. https://doi.org/10.4018/978-1-7998-4763-2.ch011

Sala, N. (2021). Virtual Reality, Augmented Reality, and Mixed Reality in Education: A Brief Overview. In D. Choi, A. Dailey-Hebert, & J. Estes (Eds.), *Current and Prospective Applications of Virtual Reality in Higher Education* (pp. 48–73). IGI Global. https://doi.org/10.4018/978-1-7998-4960-5.ch003

Salunkhe, S., Kanagachidambaresan, G., Rajkumar, C., & Jayanthi, K. (2022). Online Detection and Prediction of Fused Deposition Modelled Parts Using Artificial Intelligence. In S. Salunkhe, H. Hussein, & J. Davim (Eds.), *Applications of Artificial Intelligence in Additive Manufacturing* (pp. 194–209). IGI Global. https://doi.org/10.4018/978-1-7998-8516-0.ch009

Samy, V. S., Pramanick, K., Thenkanidiyoor, V., & Victor, J. (2021). Data Analysis and Visualization in Python for Polar Meteorological Data. *International Journal of Data Analytics, 2*(1), 32–60. https://doi.org/10.4018/IJDA.2021010102

Sanna, A., & Valpreda, F. (2017). An Assessment of the Impact of a Collaborative Didactic Approach and Students' Background in Teaching Computer Animation. *International Journal of Information and Communication Technology Education, 13*(4), 1–16. doi:10.4018/IJICTE.2017100101

Sarivougioukas, J., & Vagelatos, A. (2022). Fused Contextual Data With Threading Technology to Accelerate Processing in Home UbiHealth. *International Journal of Software Science and Computational Intelligence, 14*(1), 1–14. https://doi.org/10.4018/IJSSCI.285590

Scott, A., Martin, A., & McAlear, F. (2017). Enhancing Participation in Computer Science among Girls of Color: An Examination of a Preparatory AP Computer Science Intervention. In Y. Rankin & J. Thomas (Eds.), *Moving Students of Color from Consumers to Producers of Technology* (pp. 62–84). Hershey, PA: IGI Global. doi:10.4018/978-1-5225-2005-4.ch004

Shanmugam, M., Ibrahim, N., Gorment, N. Z., Sugu, R., Dandarawi, T. N., & Ahmad, N. A. (2022). Towards an Integrated Omni-Channel Strategy Framework for Improved Customer Interaction. In P. Lai (Ed.), *Handbook of Research on Social Impacts of E-Payment and Blockchain Technology* (pp. 409–427). IGI Global. https://doi.org/10.4018/978-1-7998-9035-5.ch022

Sharma, A., & Kumar, S. (2021). Network Slicing and the Role of 5G in IoT Applications. In S. Kumar, M. Trivedi, P. Ranjan, & A. Punhani (Eds.), *Evolution of Software-Defined Networking Foundations for IoT and 5G Mobile Networks* (pp. 172–190). IGI Global. https://doi.org/10.4018/978-1-7998-4685-7.ch010

Siddoo, V., & Wongsai, N. (2017). Factors Influencing the Adoption of ISO/IEC 29110 in Thai Government Projects: A Case Study. *International Journal of Information Technologies and Systems Approach, 10*(1), 22–44. doi:10.4018/IJITSA.2017010102

Silveira, C., Hir, M. E., & Chaves, H. K. (2022). An Approach to Information Management as a Subsidy of Global Health Actions: A Case Study of Big Data in Health for Dengue, Zika, and Chikungunya. In J. Lima de Magalhães, Z. Hartz, G. Jamil, H. Silveira, & L. Jamil (Eds.), *Handbook of Research on Essential Information Approaches to Aiding Global Health in the One Health Context* (pp. 219–234). IGI Global. https://doi.org/10.4018/978-1-7998-8011-0.ch012

Simões, A. (2017). Using Game Frameworks to Teach Computer Programming. In R. Alexandre Peixoto de Queirós & M. Pinto (Eds.), *Gamification-Based E-Learning Strategies for Computer Programming Education* (pp. 221–236). Hershey, PA: IGI Global. doi:10.4018/978-1-5225-1034-5.ch010

Simões de Almeida, R., & da Silva, T. (2022). AI Chatbots in Mental Health: Are We There Yet? In A. Marques & R. Queirós (Eds.), *Digital Therapies in Psychosocial Rehabilitation and Mental Health* (pp. 226–243). IGI Global. https://doi.org/10.4018/978-1-7998-8634-1.ch011

Singh, L. K., Khanna, M., Thawkar, S., & Gopal, J. (2021). Robustness for Authentication of the Human Using Face, Ear, and Gait Multimodal Biometric System. *International Journal of Information System Modeling and Design, 12*(1), 39–72. https://doi.org/10.4018/IJISMD.2021010103

Sllame, A. M. (2017). Integrating LAB Work With Classes in Computer Network Courses. In H. Alphin Jr, R. Chan, & J. Lavine (Eds.), *The Future of Accessibility in International Higher Education* (pp. 253–275). Hershey, PA: IGI Global. doi:10.4018/978-1-5225-2560-8.ch015

Smirnov, A., Ponomarev, A., Shilov, N., Kashevnik, A., & Teslya, N. (2018). Ontology-Based Human-Computer Cloud for Decision Support: Architecture and Applications in Tourism. *International Journal of Embedded and Real-Time Communication Systems, 9*(1), 1–19. doi:10.4018/IJERTCS.2018010101

Smith-Ditizio, A. A., & Smith, A. D. (2018). Computer Fraud Challenges and Its Legal Implications. In M. Khosrow-Pour, D.B.A. (Ed.), Encyclopedia of Information Science and Technology, Fourth Edition (pp. 4837-4848). Hershey, PA: IGI Global. doi:10.4018/978-1-5225-2255-3.ch419

Sosnin, P. (2018). Figuratively Semantic Support of Human-Computer Interactions. In *Experience-Based Human-Computer Interactions: Emerging Research and Opportunities* (pp. 244–272). Hershey, PA: IGI Global. doi:10.4018/978-1-5225-2987-3.ch008

Srilakshmi, R., & Jaya Bhaskar, M. (2021). An Adaptable Secure Scheme in Mobile Ad hoc Network to Protect the Communication Channel From Malicious Behaviours. *International Journal of Information Technology and Web Engineering, 16*(3), 54–73. https://doi.org/10.4018/IJITWE.2021070104

Sukhwani, N., Kagita, V. R., Kumar, V., & Panda, S. K. (2021). Efficient Computation of Top-K Skyline Objects in Data Set With Uncertain Preferences. *International Journal of Data Warehousing and Mining, 17*(3), 68–80. https://doi.org/10.4018/IJDWM.2021070104

Susanto, H., Yie, L. F., Setiana, D., Asih, Y., Yoganingrum, A., Riyanto, S., & Saputra, F. A. (2021). Digital Ecosystem Security Issues for Organizations and Governments: Digital Ethics and Privacy. In Z. Mahmood (Ed.), *Web 2.0 and Cloud Technologies for Implementing Connected Government* (pp. 204–228). IGI Global. https://doi.org/10.4018/978-1-7998-4570-6.ch010

Syväjärvi, A., Leinonen, J., Kivivirta, V., & Kesti, M. (2017). The Latitude of Information Management in Local Government: Views of Local Government Managers. *International Journal of Electronic Government Research, 13*(1), 69–85. doi:10.4018/IJEGR.2017010105

Tanque, M., & Foxwell, H. J. (2018). Big Data and Cloud Computing: A Review of Supply Chain Capabilities and Challenges. In A. Prasad (Ed.), *Exploring the Convergence of Big Data and the Internet of Things* (pp. 1–28). Hershey, PA: IGI Global. doi:10.4018/978-1-5225-2947-7.ch001

Teixeira, A., Gomes, A., & Orvalho, J. G. (2017). Auditory Feedback in a Computer Game for Blind People. In T. Issa, P. Kommers, T. Issa, P. Isaías, & T. Issa (Eds.), *Smart Technology Applications in Business Environments* (pp. 134–158). Hershey, PA: IGI Global. doi:10.4018/978-1-5225-2492-2.ch007

Tewari, P., Tiwari, P., & Goel, R. (2022). Information Technology in Supply Chain Management. In V. Garg & R. Goel (Eds.), *Handbook of Research on Innovative Management Using AI in Industry 5.0* (pp. 165–178). IGI Global. https://doi.org/10.4018/978-1-7998-8497-2.ch011

Thompson, N., McGill, T., & Murray, D. (2018). Affect-Sensitive Computer Systems. In M. Khosrow-Pour, D.B.A. (Ed.), Encyclopedia of Information Science and Technology, Fourth Edition (pp. 4124-4135). Hershey, PA: IGI Global. doi:10.4018/978-1-5225-2255-3.ch357

Triberti, S., Brivio, E., & Galimberti, C. (2018). On Social Presence: Theories, Methodologies, and Guidelines for the Innovative Contexts of Computer-Mediated Learning. In M. Marmon (Ed.), *Enhancing Social Presence in Online Learning Environments* (pp. 20–41). Hershey, PA: IGI Global. doi:10.4018/978-1-5225-3229-3.ch002

Tripathy, B. K. T. R., S., & Mohanty, R. K. (2018). Memetic Algorithms and Their Applications in Computer Science. In S. Dash, B. Tripathy, & A. Rahman (Eds.), Handbook of Research on Modeling, Analysis, and Application of Nature-Inspired Metaheuristic Algorithms (pp. 73-93). Hershey, PA: IGI Global. https://doi.org/ doi:10.4018/978-1-5225-2857-9.ch004

Turulja, L., & Bajgoric, N. (2017). Human Resource Management IT and Global Economy Perspective: Global Human Resource Information Systems. In M. Khosrow-Pour (Ed.), *Handbook of Research on Technology Adoption, Social Policy, and Global Integration* (pp. 377–394). Hershey, PA: IGI Global. doi:10.4018/978-1-5225-2668-1.ch018

Unwin, D. W., Sanzogni, L., & Sandhu, K. (2017). Developing and Measuring the Business Case for Health Information Technology. In K. Moahi, K. Bwalya, & P. Sebina (Eds.), *Health Information Systems and the Advancement of Medical Practice in Developing Countries* (pp. 262–290). Hershey, PA: IGI Global. doi:10.4018/978-1-5225-2262-1.ch015

Usharani, B. (2022). House Plant Leaf Disease Detection and Classification Using Machine Learning. In M. Mundada, S. Seema, S. K.G., & M. Shilpa (Eds.), *Deep Learning Applications for Cyber-Physical Systems* (pp. 17-26). IGI Global. https://doi.org/10.4018/978-1-7998-8161-2.ch002

Vadhanam, B. R. S., M., Sugumaran, V., V., V., & Ramalingam, V. V. (2017). Computer Vision Based Classification on Commercial Videos. In M. S., & V. V. (Eds.), Multi-Core Computer Vision and Image Processing for Intelligent Applications (pp. 105-135). Hershey, PA: IGI Global. https://doi.org/ doi:10.4018/978-1-5225-0889-2.ch004

Vairinho, S. (2022). Innovation Dynamics Through the Encouragement of Knowledge Spin-Off From Touristic Destinations. In C. Ramos, S. Quinteiro, & A. Gonçalves (Eds.), *ICT as Innovator Between Tourism and Culture* (pp. 170–190). IGI Global. https://doi.org/10.4018/978-1-7998-8165-0.ch011

Valverde, R., Torres, B., & Motaghi, H. (2018). A Quantum NeuroIS Data Analytics Architecture for the Usability Evaluation of Learning Management Systems. In S. Bhattacharyya (Ed.), *Quantum-Inspired Intelligent Systems for Multimedia Data Analysis* (pp. 277–299). Hershey, PA: IGI Global. doi:10.4018/978-1-5225-5219-2.ch009

Vassilis, E. (2018). Learning and Teaching Methodology: "1:1 Educational Computing. In K. Koutsopoulos, K. Doukas, & Y. Kotsanis (Eds.), *Handbook of Research on Educational Design and Cloud Computing in Modern Classroom Settings* (pp. 122–155). Hershey, PA: IGI Global. doi:10.4018/978-1-5225-3053-4.ch007

Verma, S., & Jain, A. K. (2022). A Survey on Sentiment Analysis Techniques for Twitter. In B. Gupta, D. Peraković, A. Abd El-Latif, & D. Gupta (Eds.), *Data Mining Approaches for Big Data and Sentiment Analysis in Social Media* (pp. 57–90). IGI Global. https://doi.org/10.4018/978-1-7998-8413-2.ch003

Wang, H., Huang, P., & Chen, X. (2021). Research and Application of a Multidimensional Association Rules Mining Method Based on OLAP. *International Journal of Information Technology and Web Engineering, 16*(1), 75–94. https://doi.org/10.4018/IJITWE.2021010104

Wexler, B. E. (2017). Computer-Presented and Physical Brain-Training Exercises for School Children: Improving Executive Functions and Learning. In B. Dubbels (Ed.), *Transforming Gaming and Computer Simulation Technologies across Industries* (pp. 206–224). Hershey, PA: IGI Global. doi:10.4018/978-1-5225-1817-4.ch012

Wimble, M., Singh, H., & Phillips, B. (2018). Understanding Cross-Level Interactions of Firm-Level Information Technology and Industry Environment: A Multilevel Model of Business Value. *Information Resources Management Journal, 31*(1), 1–20. doi:10.4018/IRMJ.2018010101

Wimmer, H., Powell, L., Kilgus, L., & Force, C. (2017). Improving Course Assessment via Web-based Homework. *International Journal of Online Pedagogy and Course Design, 7*(2), 1–19. doi:10.4018/IJOPCD.2017040101

Wong, S. (2021). Gendering Information and Communication Technologies in Climate Change. In M. Khosrow-Pour D.B.A. (Eds.), *Encyclopedia of Information Science and Technology, Fifth Edition* (pp. 1408-1422). IGI Global. https://doi.org/10.4018/978-1-7998-3479-3.ch096

Wong, Y. L., & Siu, K. W. (2018). Assessing Computer-Aided Design Skills. In M. Khosrow-Pour, D.B.A. (Ed.), Encyclopedia of Information Science and Technology, Fourth Edition (pp. 7382-7391). Hershey, PA: IGI Global. doi:10.4018/978-1-5225-2255-3.ch642

Wongsurawat, W., & Shrestha, V. (2018). Information Technology, Globalization, and Local Conditions: Implications for Entrepreneurs in Southeast Asia. In P. Ordóñez de Pablos (Ed.), *Management Strategies and Technology Fluidity in the Asian Business Sector* (pp. 163–176). Hershey, PA: IGI Global. doi:10.4018/978-1-5225-4056-4.ch010

Yamada, H. (2021). Homogenization of Japanese Industrial Technology From the Perspective of R&D Expenses. *International Journal of Systems and Service-Oriented Engineering, 11*(2), 24–51. doi:10.4018/IJSSOE.2021070102

Yang, Y., Zhu, X., Jin, C., & Li, J. J. (2018). Reforming Classroom Education Through a QQ Group: A Pilot Experiment at a Primary School in Shanghai. In H. Spires (Ed.), *Digital Transformation and Innovation in Chinese Education* (pp. 211–231). Hershey, PA: IGI Global. doi:10.4018/978-1-5225-2924-8.ch012

Yilmaz, R., Sezgin, A., Kurnaz, S., & Arslan, Y. Z. (2018). Object-Oriented Programming in Computer Science. In M. Khosrow-Pour, D.B.A. (Ed.), Encyclopedia of Information Science and Technology, Fourth Edition (pp. 7470-7480). Hershey, PA: IGI Global. doi:10.4018/978-1-5225-2255-3.ch650

Yu, L. (2018). From Teaching Software Engineering Locally and Globally to Devising an Internationalized Computer Science Curriculum. In S. Dikli, B. Etheridge, & R. Rawls (Eds.), *Curriculum Internationalization and the Future of Education* (pp. 293–320). Hershey, PA: IGI Global. doi:10.4018/978-1-5225-2791-6.ch016

Yuhua, F. (2018). Computer Information Library Clusters. In M. Khosrow-Pour, D.B.A. (Ed.), Encyclopedia of Information Science and Technology, Fourth Edition (pp. 4399-4403). Hershey, PA: IGI Global. doi:10.4018/978-1-5225-2255-3.ch382

Zakaria, R. B., Zainuddin, M. N., & Mohamad, A. H. (2022). Distilling Blockchain: Complexity, Barriers, and Opportunities. In P. Lai (Ed.), *Handbook of Research on Social Impacts of E-Payment and Blockchain Technology* (pp. 89–114). IGI Global. https://doi.org/10.4018/978-1-7998-9035-5.ch007

Zhang, Z., Ma, J., & Cui, X. (2021). Genetic Algorithm With Three-Dimensional Population Dominance Strategy for University Course Timetabling Problem. *International Journal of Grid and High Performance Computing, 13*(2), 56–69. https://doi.org/10.4018/IJGHPC.2021040104

About the Contributors

Fadi Al-Turjman received his Ph.D. in computer science from Queen's University, Canada, in 2011. He is the "advisor to the chairman of the board of trustees on AI and informatics" at Near East University (NEU), Nicosia, and the "associate dean for research" at the faculty of engineering in the same university. Prof. Al-Turjman is the head of Artificial Intelligence (AI), Information Systems, and Software Engineering Departements, and the director for the AI and Robotics Institute and the International Research Center for AI and IoT at NEU. He has been awarded the lifetime golden-award of Dr. Suat Gunsel from Near East University for the year 2022, in addition to several other recognitions and best research awards at key international venues. Prof. Al-Turjman is a leading authority in the areas of smart/intelligent IoT systems, wireless, and mobile networks' architectures, protocols, deployments, and performance evaluation in Artificial Intelligence of Things (AIoT). His publication history spans over 600 SCI/E publications, in addition to numerous keynotes and plenary talks at flagship venues. He has authored and edited more than 50 books about AI, cognition, security, and wireless sensor networks' deployments in smart IoT environments, which have been published by well-reputed publishers such as IEEE, Taylor and Francis, Elsevier, IET, and Springer. Prof. Al-Turjman has led a number of international conferences and workshops in flagship AI and IoT societies. Currently, he serves as a book series editor and the lead guest/associate editor for several top tier journals, including the IEEE Communications Surveys and Tutorials (IF 23.9) and the Elsevier Sustainable Cities and Society (IF 10.8), in addition to organizing international conferences and symposiums on the most up to date research topics in AI and IoT.

* * *

Neepa Biswas has done her B.Tech in Information Technology from WBUT, M.E. in Information Technology from IIEST Shibpur, She has completed PhD from Jadavpur University. Her research Interest Data Warehousing, Data mining, Ad-hoc

Network, Machine Learning etc. Currently she is working as Assistant Professor in Narula Institute of Technology, in West Bengal, Kolkata, India.

Sudarsan Biswas completed his Bachelor of Engineering (IT) from Bengal Engineering and Science University, Shibpur in 2006 and his Master of Engineering (IT) from Bengal Engineering and Science University, Shibpur in 2010. Currently serving as an Assistant Professor Department of IT, RCC Institute of Information Technology, Kolkata, West Bengal, India.

Dankan Gowda V. is currently working as an Assistant Professor in the Department of Electronics and Communication Engineering at BMS Institute of Technology and Management in Bangalore. Previously, he worked as a Research Fellow at ADA DRDO and as a Software Engineer at Robert Bosch. With a total experience of 14 years, including teaching and industry, Dr. Gowda has made significant contributions to both academia and research. He has published over 60 research papers in renowned international journals and conferences. In recognition of his innovative work, Dr. Gowda has been granted six patents, including four from Indian authorities and two international patents. His research interests primarily lie in the fields of IoT and Signal Processing, where he has conducted workshops and handled industry projects. Dr. Gowda is passionate about teaching and strives to create an engaging and effective learning environment for his students. He consistently seeks opportunities to enhance his knowledge and stay updated with the latest advancements in his field.

Manish Jain obtained B.E. in 2001 in Electronics and Telecommunication Engg., M.Tech in 2008 in Microelectronics & VLSI design & Ph.D. in 2015 in Electronics & Communication. He has a rich teaching experience of 17 years & worked in various prestigious Engg. Colleges. He has 30 papers published in international journal and some listed in IEEE conference Proceedings. He conducted couples of Conferences, workshop and FDP, He is life member of IE, CEGR, IAENG and Editorial board members &Reviewer of some International Journal. He has authored 3 books on DSP, VLSI field. He has filed 02 patents. Under his guidance Student project have been sanctioned for Financial Assistance by Department of Science & Technology (DST). Under his guidance 10 M.Tech and 1 Ph.d has been awarded. His research interest includes Semiconductor Electronics, Communication, ADC design & Testing, VLSI.

Mandeep Kaur is working as Assistant Professor in Department of Electronics and Communication engineering, Punjabi University Patiala. She has thirteen years of teaching experience. She has published more than 80 research papers in SCI and

Scopus journal and International/National conferences. She has guided more than 25 M.Tech students. Her area of interest is Digital signal Processing, Artificial intelligence, optimization and machine learning, IoT.

Rupesh Kumar was born in Bokaro Steel City, India, in 1983. He received B.Tech degree in Electronics & Communication engineering from SMIT, India, in 2008, and MS and PhD degrees in Electronics & Communication from Institut Mines-Telecom, Paris, France, in 2011 and 2014. Currently, he is working as a Professor at SSRM AP University, India. His research is focused on indoor localization system, FMCW-CW Radars, and Reconfigurable Array Antenna, RFID Technology, Wireless Sensing, Machine Learning.

Yogesh Patidar received the B.E. degree from the Government Engineering College Ujjain, MP, India, in 2004, and the MTech. degree from the RKDF University, Bhopal, MP, India, in 2018. He has 17 years of teaching experience. He is currently pursuing the Ph.D. degree with the Department of Electrical and Electronics Engineering, Mandsaur University, Mandsaur, MP, India. His current research interest includes wireless sensor network, cognitive radio, digital communication and signal processing.

K. D. V. Prasad works as a Faculty of Research at Symbiosis Institute of Business Management, Hyderabad; Symbiosis International (Deemed University), Pune, India. Dr. Prasad holds a Master's in Computer Applications and a Master's in Software Systems from BITS, Pilani; MBA (Human Resources), IGNOU, New Delhi. Dr. Prasad possesses a Ph.D. in Business Management (Kanpur University), and PhD in Business Administration (RTM Nagpur University. He is AIMA-Certified Management Teacher and Fellow, World HR Board, Carlton Advanced Management Institute, USA; Fellowship of the Management and Business Research Council (FMBRC) by Open Association Research Society, Delaware, USA Dr. Prasad Published over 100 articles in Scopus/WoS indexed journals and 3 books.

Shekhar R. is a Doctorate in Computer Science from MS University, Tirunelveli, a Postgraduate in Computer Networks and graduate in Computer Science and Engineering from the University of Bangalore. He has professional certifications in Microsoft, Cisco Networks. Dr. Shekhar joined Alliance in 2012 and has since been actively engaged in the areas of teaching, research, training, and development. He has vast experience of more than two decades in teaching and has held many prestigious positions in various reputed institutions. Dr. Shekhar has been awarded the " Rastriya Shiksha Gaurav Puraskar" a National award in the year 2021, "Techno Guru" a National Award in 2019, "Adarsha Vidya Saraswathi Rashtriya Puraskar" a National award and "Distinguished Faculty" from reputed organizations in the year 2018. Dr. Shekhar has been awarded the 'Best Editor' for contributing as 'Lead Editor' for the book 'Maximum Authors Contributing For A Book on Covid 19 and Its Impact' registered for the event 'India book of Records' and 'Asia book of Records'. Dr. Shekhar has an Indian Patent Published on Artificial intelligence. He has authored for Textbooks related to Computer Science and Engineering-Introduction to Microprocessor, Nano-computing and Cyber heal. He is a lifetime member of ISTE, CSI, IET holds membership of various reputed international organizations. He serves as an Editorial member of IJICTS, an international journal and has been a reviewer for several National and International journals. He is also a sitting member of different National and International Associations in the field of Computer science. Dr. Shekhar has been working as SPOC for SWAYAM NPTEL since 2018 and has received appreciation and has been awarded as Active SPOC for SWAYAM-NPTEL. Dr. Shekhar has been teaching for Post-Graduation, UnderGraduation and PhD students in the field of Computer Science and Engineering He is a highly versatile teacher in various Computer related subjects such as data Mining, Image Processing, Computer Networks, Artificial Intelligence, and many other subjects related to Computer Science and Engineering. He has guided many postgraduate students in preparing dissertations in various subjects. He has presented papers in National and International forums, published number of articles in journals having high impact factors, and chaired many conferences. He has organized various workshops and conferences for highly reputed institutions.

Debarpita Santra has completed her M.Sc. in Computer Science from Banaras Hindu University, and M.Tech in Computer Science & Engineering from University of Calcutta. She has done her Ph.D. thesis from University of Kalyani. She has many research articles published in International Journals, conferences, and books. Her broad research interest is in Artificial Intelligence in Healthcare.

D. Srinivas distinguished Faculty of Marketing and Communication at SR University, Warangal, Telangana. He received MBA from Kakatiya University, Warangal and PhD degree from Kakatiya University, Warangal. He is the author of many books. He also published many articles in National and International journals. His research covers Product strategy, Digital marketing, E-Business and IoT For Business Operations. He is a active member of AIMA and HMA. He has delivered many guest lectures and organized FDP's across the country.

Raj Gaurang Tiwari received Ph.D. (Computer Science) in 2013, M. Tech. (Computer Science and Engineering) in 2010 and Master in Computer Application in 2002. He has teaching and research experience of 20 years. He is presently working as Professor and Dean in the Department of Computer Science and Engineering at Chitkara University, Punjab. He also worked as Associate Professor in the Department of Computer Science and Engineering at Sri Ramswaroop Memorial Group of Professional Colleges, Lucknow and as Assistant Professor at AZAD Institute of Engineering and Technology, Lucknow. His research interests primarily embrace Knowledge-Based Engineering, Web Engineering, Ad-hoc/Sensor Network, Recommender Systems and Data Science. He has authored more than 70 International and national journal and conference papers. He is member of Programme Committee / Technical Committee /Reviewer for many international conferences/journals.

Ajay Kumar Vyas is currently working at Department of Information and Communication Technology, Adani University, Ahmedabad, India. He is a fellow member of IETE, a senior member of IEEE, a life member of ISTE. He has completed his Bachelor of Engineering in Electronics & Communication for GEC Ujjain and M.Tech for Shri GSITS Indore in field of optical communication with Honors. He earned his Ph.D. from MPUAT, Udaipur, India. He has the author of numerous research documents in indexed journals and conferences. He is a reviewer and a member of the organizing committees of several international congresses in the Optical Communication and WSN.

Index

N

P

R

S

T

V

W

X

www.ingramcontent.com/pod-product-compliance
Lightning Source LLC
LaVergne TN
LVHW081324300926
849161LV00008B/1872